21世纪高职高专IT类专业系列教材
SHIJI GAOZHI GAOZHUAN IT LEI ZHUANYE XILIE JIAOCAI

Yunjisuan Jiagou yu Yingyong

云计算架构与应用

主　编　朱义勇
副主编　周向军　陈志涛　张宇辉　曹继翔

U0396512

·广州·

图书在版编目（CIP）数据

云计算架构与应用/朱义勇主编. —广州：华南理工大学出版社，2017.8（2019.1 重印）
21 世纪高职高专 IT 类专业系列教材
ISBN 978-7-5623-5353-9

Ⅰ.①云⋯　Ⅱ.①朱⋯　Ⅲ.①云计算－高等职业教育－教材　Ⅳ.①TP393.027

中国版本图书馆 CIP 数据核字（2017）第 194739 号

云计算架构与应用
朱义勇　主编

出 版 人：**卢家明**
出版发行：华南理工大学出版社
　　　　　（广州五山华南理工大学 17 号楼，邮编 510640）
　　　　　http：//www.scutpress.com.cn　　E-mail：scutc13@scut.edu.cn
　　　　　营销部电话：020-87113487　87111048（传真）
责任编辑：何丽云
特约编辑：蔡贤资
印　刷　者：虎彩印艺股份有限公司
开　　　本：787mm×960mm　1/16　印张：15.25　字数：352 千
版　　　次：2017 年 8 月第 1 版　2019 年 1 月第 3 次印刷
定　　　价：39.80 元

版权所有　盗版必究　印装差错　负责调换

序

  云计算被认为是继个人电脑、互联网之后信息技术的又一次重大变革,将带来工作方式和商业模式的根本性改变。近年来,我国政府高度重视云计算产业发展,国务院先后出台了一系列与云计算密切相关的政策文件推动云计算产业发展、行业推广、技术应用、安全管理等。同时教育部在2015年将云计算应用与技术纳入到高等院校学科设置专业目录,云计算相关技术的理论和实践都有了飞速的发展。

  随着云计算相关技术的广泛应用,无论是高校的云计算技术与应用专业人才培养,还是其他人才培训机构,云计算相关技能人才培养工作都开展得如火如荼,我们可以预见今后几年内会有越来越多的技术人员参与到云计算相关领域工作中来。云平台作为云计算应用的重要组成部分,在实践中得到广泛的应用。目前云平台分为开源和非开源两种形式,开源云平台主要以 OpenStack 和 CloudStack 为主,均支持 VMWare、KVM、Hyper-V 以及 Xen 等各种虚拟化技术,满足各种云计算虚拟化底层的要求。在开源的云平台学习资源方面,互联网提供了丰富的技术资料,加上各类云计算专业技术类书籍,为广大的云计算专业技术人员以及爱好者提供了较多选择。而市场上多数云计算相关教材偏向技术理论方面,缺乏与实际应用操作相结合。

  为加强技能应用型人才培养力度,本书作者凭借多年的企业实践以及十余年的教学研究经验,联合企业共同开发,在编写过程中力求理论与实践操作相结合,深入浅出地描述开源 OpenStack 云计算管理平台服务组件的技术理论,并通过广东时汇信息科技有限公司的蜜蜂实训平台创建实例应用,为读者提供了详细的安装部署过程,是一本理论与实践相结合的教学指导书籍。

希望本书的读者,在学习开源 OpenStack 云计算管理平台应用的同时,加强云计算实践与应用,只有更多的人认识到云计算的价值,充分挖掘云计算的价值,云计算产业才会形成源源不断的动力,推动云计算产业的发展。

2017 年 6 月

(韩国强系华南理工大学计算机科学与工程学院院长、教授,教育部计算机教学指导委员会委员,广东省计算机学会会长)

前言

OpenStack 是一个开源的云计算管理平台项目，最初由 NASA 和 Rackspace 合作研发并发起。OpenStack 项目发展到今天已成为开源云计算技术中的引领者。

OpenStack 是一个旨在为公有云及私有云的建设与管理提供软件服务的开源项目，吸引了众多开发者以及知名厂商参与其中。OpenStack 为用户提供了基础设施即服务（IaaS）的解决方案。

本书总计15章，以 OpenStack 为基础来阐述云计算架构及其应用。从认识云计算和 OpenStack 项目，到构建一个功能完整的 OpenStack 云平台，最后创建实例应用，为读者提供了详细的安装部署过程指导，同时还讲述了各 OpenStack 服务组件的技术理论。本书涵盖了 OpenStack 项目主要的服务组件，包括 Nova（计算服务）、Glance（镜像服务）、Keystone（认证服务）、Neutron（网络服务）、Horizon（仪表板服务）、Cinder（块存储服务）、Manila（文件共享服务）、Swift（对象存储服务）、Ceilometer（计量服务）和 Heat（编排服务）。

本书适合初次接触 OpenStack 的读者学习，通过本书的学习，读者可以轻松进入 OpenStack 的世界，并学以致用。

本书由朱义勇担任主编，周向军、陈志涛、张宇辉、曹继翔任副主编。朱义勇编写第1、2、3、4、5、6、7章并负责本书的统筹统稿工作，陈志涛编写第8、9章，张宇辉编写第10、11、12章，曹继翔编写第13、14章，周向军编写第15章。

本书在写作过程中得到了广东时汇信息科技有限公司大力支持，并使用了其公司提供的蜜蜂实训平台，为实战操作提供了可靠的应用环境，同时公司协调周剑辉、周政江等云计算工程师对实验步骤进行了详细的验证，在此表示衷心感谢！

编 者

2017年4月

目 录

第 1 章　云计算概述 ·· (1)
  1.1　云计算的定义 ·· (1)
  1.2　云计算的基本特征 ·· (1)
  1.3　云计算的服务模型 ·· (2)
  1.4　云计算的发布模型 ·· (4)
  1.5　云计算的平台分类 ·· (5)

第 2 章　初识 OpenStack ·· (7)
  2.1　OpenStack 发展历程 ·· (7)
  2.2　OpenStack 概述 ··· (7)
  2.3　OpenStack 架构 ··· (8)
  2.4　OpenStack 服务组件 ··· (11)
  2.5　OpenStack 版本演变 ··· (13)

第 3 章　OpenStack 架构示例 ··· (16)
  3.1　部署架构 ··· (16)
  3.2　硬件配置 ··· (16)
  3.3　服务组件 ··· (17)
  3.4　虚拟网络类型 ··· (18)
  3.5　主机网络 ··· (18)
  3.6　密码配置 ··· (19)

第 4 章　准备 OpenStack 初始环境 ·· (21)
  4.1　操作系统 ··· (21)
  4.2　进入终端 ··· (22)
  4.3　关闭防火墙（所有节点） ··· (24)
  4.4　配置主机名（所有节点） ··· (24)

I

4.5　配置网络（所有节点） …………………………………………………………（25）
4.6　配置 DNS 解析（所有节点） ……………………………………………………（30）
4.7　配置本地安装源（所有节点） ……………………………………………………（30）
4.8　配置 NTP（所有节点） …………………………………………………………（33）
4.9　启用 OpenStack 库（所有节点） …………………………………………………（35）
4.10　更新操作系统（所有节点） ……………………………………………………（35）
4.11　安装 OpenStack 客户端（所有节点） …………………………………………（36）
4.12　安装 SQL 数据库（控制节点） …………………………………………………（36）
4.13　安装消息队列 RabbitMQ（控制节点） …………………………………………（38）
4.14　安装缓存服务 Memcached（控制节点） ………………………………………（38）

第 5 章　身份认证（Keystone）服务 …………………………………………………（39）
5.1　服务概述 ……………………………………………………………………………（39）
5.2　安装前准备（控制节点） …………………………………………………………（42）
5.3　安装和配置（控制节点） …………………………………………………………（42）
5.4　创建服务实体和 API 端点（控制节点） …………………………………………（45）
5.5　创建域、项目、用户和角色（控制节点） ………………………………………（48）
5.6　验证操作（控制节点） ……………………………………………………………（51）
5.7　创建 OpenStack 客户端环境脚本（控制节点） …………………………………（53）

第 6 章　镜像（Glance）服务 …………………………………………………………（55）
6.1　服务概述 ……………………………………………………………………………（55）
6.2　安装前准备（控制节点） …………………………………………………………（56）
6.3　安装和配置（控制节点） …………………………………………………………（59）
6.4　验证操作（控制节点） ……………………………………………………………（62）

第 7 章　计算（Nova）服务 ……………………………………………………………（64）
7.1　服务概述 ……………………………………………………………………………（64）
7.2　安装前准备（控制节点） …………………………………………………………（66）
7.3　安装和配置（控制节点） …………………………………………………………（69）
7.4　安装和配置（计算节点） …………………………………………………………（73）
7.5　验证操作（控制节点） ……………………………………………………………（76）

第 8 章 网络（Neutron）服务 (78)
8.1 服务概述 (78)
8.2 安装前准备（控制节点） (80)
8.3 安装和配置（控制节点） (83)
8.4 安装和配置（计算节点） (92)
8.5 验证操作（控制节点） (96)

第 9 章 仪表板（Horizon）服务 (99)
9.1 服务概述 (99)
9.2 安装和配置（控制节点） (99)
9.3 验证操作（控制节点） (101)

第 10 章 块存储（Cinder）服务 (105)
10.1 服务概述 (105)
10.2 安装前准备（控制节点） (106)
10.3 安装和配置（控制节点） (110)
10.4 安装和配置（块存储节点） (113)
10.5 验证操作（控制节点） (117)

第 11 章 文件共享（Manila）服务 (118)
11.1 服务概述 (118)
11.2 安装前准备（控制节点） (119)
11.3 安装和配置（控制节点） (123)
11.4 安装和配置（块存储节点） (126)
11.5 验证操作（控制节点） (132)

第 12 章 对象存储（Swift）服务 (134)
12.1 服务概述 (134)
12.2 安装前准备（控制节点） (135)
12.3 安装和配置（控制节点） (138)
12.4 安装和配置（对象存储节点） (140)
12.5 创建、分发并初始化 rings（控制节点） (146)
12.6 完成安装（控制节点、对象存储节点） (151)
12.7 验证操作（控制节点） (153)

第 13 章 编排（Heat）服务 (156)
13.1 服务概述 (156)
13.2 安装前准备（控制节点） (156)
13.3 安装和配置（控制节点） (163)
13.4 验证操作（控制节点） (166)

第 14 章 计量（Ceilometer）服务 (167)
14.1 服务概述 (167)
14.2 安装前准备（控制节点） (168)
14.3 安装和配置（控制节点） (172)
14.4 启用镜像服务计量（控制节点） (174)
14.5 启用计算服务计量（计算节点） (175)
14.6 启用块存储计量（控制节点、块存储节点） (177)
14.7 启用对象存储服务计量（控制节点） (178)
14.8 安装计量警告服务（控制节点） (180)
14.9 验证操作（控制节点） (185)

第 15 章 创建虚拟机实例 (188)
15.1 创建虚拟网络（控制节点） (188)
15.2 创建 m1.nano 规格主机（控制节点） (199)
15.3 生成密钥对（控制节点） (200)
15.4 为 default 安全组添加规则（控制节点） (201)
15.5 创建虚拟机实例（控制节点） (201)
15.6 创建块设备存储（控制节点） (212)
15.7 创建编排（控制节点） (214)
15.8 访问仪表板（控制节点） (218)

参考文献 (233)

第 1 章　云计算概述

1.1　云计算的定义

1961 年，John McCarthy 公开提出了"云"中计算的概念："如果我倡导的计算机能在未来得到使用，那么有一天，计算也可能像电话一样成为公用设施。……计算机应用（Computer utility）将成为一种全新的、重要的产业基础。"

虽然云计算的定义有很多种，但业内基本认可美国国家标准与技术研究院（NIST）对云计算的概念定义："所谓云计算，就是一种允许用户通过无所不在的、便捷的、按需获得的网络接入到一个可动态配置的共享计算资源池（其中包括了网络设备、服务器、存储、应用以及业务），并且以最小的管理成本或者业务服务提供者交互复杂度即可实现这些可配置计算资源的快速提供与发布的模式。"

云计算的核心可以用五大基本特征、三种服务模式以及四类部署模式来概括。五大基本特征是：按需的自助服务、广泛的网络接入、资源池化、快速的弹性伸缩、可计量的服务。三种服务模式为：基础设施即服务（IaaS）、平台即服务（PaaS）、软件即服务（SaaS）。四类部署模式可以划分为：私有云、社区云、公有云、混合云。

以亚马逊 2006 年 3 月 13 日发布的在线存储服务（Simple Storage Service，S3）为起点，"云计算"这一术语开始出现在商业领域。到 2008 年 4 月 Google 发布 Google App Engine 并提出了"云计算"概念，至今已有 10 年多的历史了，其核心理念已广为人们所传播和接受，也经历了一段方兴未艾的发展过程。

1.2　云计算的基本特征

1. 按需的自助服务

客户能够单方面按需自动调配计算资源，例如服务器时间和网络存储，这些都是自动进行而无须人为干涉的。

2. 广泛的网络接入

具有通过规范机制网络访问资源的能力，这种机制可以使用各种各样的瘦和胖客户端平台（例如携带电话、笔记本电脑以及 PDA）。

3. 资源池化

提供商提供的计算资源被集中起来通过一个多客户共享模型来为多个客户提供服务，

并根据客户的需求，动态地分配或再分配不同的物理和虚拟资源。其中有一个区域独立的观念，就是客户通常不需要控制或者需要知道被提供的资源的确切的位置，但是可能会在更高一层的抽象（例如国家、州或者数据中心）上指定资源的位置。资源的例子包括存储设备、数据加工、内存、网络带宽和虚拟机等。

4. 快速的弹性伸缩

云计算具有快速地可伸缩性地提供服务的能力。在一些场景中，所提供的服务可以自动，并快速地横向扩展，在某种条件下也能迅速释放以及快速横向收缩。对于客户来讲，这种能力使云系统所提供的服务看起来好像是无限的，并且可以在任何时间购买任何数量。

5. 可计量的服务

云系统通过一种可计量的能力杠杆在某些抽象层上自动地控制并优化资源以达到某种服务类型（例如存储、处理、带宽以及活动用户账号）的计量。资源的使用可以被监视和控制，通过向供应商和用户提供这些被使用服务报告以达到透明化。

1.3 云计算的服务模型

1. 基础设施即服务（IaaS）

基础设施即服务（Infrastructure-as-a-Service，IaaS），向用户提供处理、存储、网络以及其他基础计算资源，客户可以在其上运行任意软件，包括操作系统和应用程序，如图 1-1 所示。用户不需要管理或者控制底层的云基础架构，但是可以控制操作系统、存储、发布应用程序，以及可以有限度地控制选择的网络组件（例如，防火墙）。如 Amazon AWS、Rackspace 等。

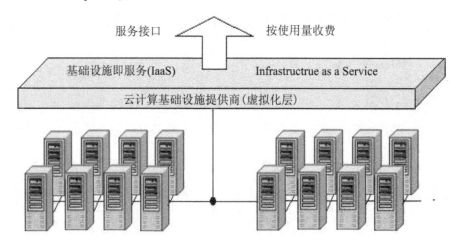

图 1-1 IaaS 模型

2. 平台即服务（PaaS）

平台即服务（Platform-as-a-Service，PaaS），用户使用云供应商支持的开发语言和工具，开发出应用程序，并发布到云基础架构上，如图 1-2 所示。用户不需要管理或者控制底层的云基础架构，包括网络、服务器、操作系统或者存储设备，但是能控制发布应用程序和可能的应用程序运行环境配置。平台通常是应用程序的基础架构，如 Google App Engine。

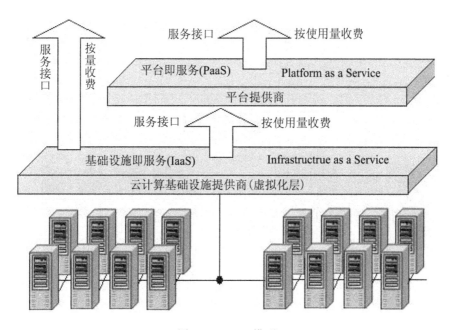

图 1-2　PaaS 模型

3. 软件即服务（SaaS）

软件即服务（Software-as-a-Service，SaaS），用户所使用的服务，是由服务商提供的运行在云基础设施上的应用程序。这些应用程序可以通过各种各样的客户端设备来访问，例如，通过客户端界面访问 WEB 浏览器的电子邮件服务。用户不需要管理或者控制底层的云基础架构，包括网络、服务器、操作系统、存储设备，甚至是独立的应用程序机能，在可能异常的情况下，会限制户可配置的应用程序设置，如图 1-3 所示。

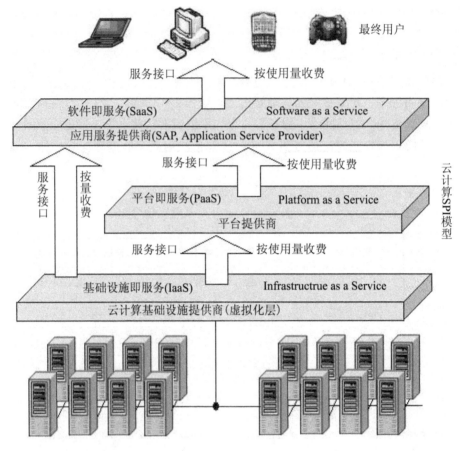

图 1-3 SaaS 模型

1.4 云计算的发布模型

1. 私有云

云基础架构被一个组织独立地操作，由这个组织或者第三方机构所管理，其核心属性是专有资源。

2. 社区云

云基础架构被几个组织所共享，并且是支持一个互相分享概念（例如任务、安全需求、策略和切合的决策）的特别社区。由这些组织或者第三方机构所管理，实现面向区域需求的服务模式。

3. 公有云

云基础架构被做成公众或者一个大的工业群体所使用，被某个组织所拥有，并出售云服务。

4. 混合云

云基础架构是由两个或者两个以上的云组成，这些云保持着唯一的实体但通过标准或者特有的技术结合在一起。这些技术使得数据或者应用程序具有可移植性（例如在云之间进行负载平衡的 Cloud Bursting 技术）。

1.5 云计算的平台分类

按云计算平台的技术应用来划分，云计算平台可以划分为 3 类：以数据存储为主的存储型云平台，以数据处理为主的计算型云平台以及计算和数据存储处理兼顾的综合云计算平台。

按构建云计算平台过程是否收费来划分，云计算平台可以划分为 2 类：开源云计算平台和商业化云计算平台。

1. 典型的开源云计算平台

1）AbiCloud（Abiquo 公司）

AbiCloud 是一款用于公司的开源云计算平台，使公司能够以快速、简单和可扩展的方式创建和管理大型、复杂的 IT 基础设施（包括虚拟服务器、网络、应用和存储设备等）。

2）Hadoop（Apache 基金会）

Hadoop 完全模仿 Google 的体系结构，是一个开源项目，主要包括 Map/Reduce 和 HDFS 文件系统。

3）Eucalyptus 项目（加利福尼亚大学）

该项目创建了一个使企业能够使用内部 IT 资源（包括服务器、存储系统、网络设备）的开源界面，进而使公司建立能够和 Amazon EC2（Elastic compute cloud，弹性云计算）兼容的云。

4）MongoDB（10gen）

MongoDB 是一个高性能、开源、无模式的文档型数据库，它在许多场景下可用于替代传统的关系型数据库或键/值存储方式，它由 C++ 语言编写。

5）OpenStack

OpenStack 是一个由 NASA（美国国家航空航天局）和 Rackspace 合作研发的，以 Apache 许可证授权的自由软件和开放源代码项目，也是一个开源的云计算管理平台项目，主要由若干个的组件组合起来完成具体工作。OpenStack 支持几乎所有类型的云环境，项目目标是提供实施简单、可大规模扩展、丰富、标准统一的云计算管理平台。OpenStack 通过各种互补的服务提供了基础设施即服务（Iaas）的解决方案，每个服务都会提供 API 以进行集成。

2. 典型的商业化云计算平台

国内典型的商业化云计算平台有阿里云、盛大云、新浪云等，这些公司都是将所拥有的云计算平台作为基础架构层的 Iaas，也即为他们所提供的云主机服务。另外，还有将云计算平台作为平台层的模式，包括腾讯的开放平台和新浪的开放平台（Pass）。他们的概念和 Google 公司的 AppEngine 相似，目的是让更多的开发者做应用，实质上这些模式都是在看到了 Apple 公司 App Store 的成功商业模式之后而形成的。相比于国外应用层的服务，国内应用层（Saas）仍是任重而道远。

国外典型的商业化云计算平台有微软、Google、IBM、Oracle、Amazon 等。这些国外云计算平台主要提供云企业服务。如微软公司的 Azure 平台；Google 公司的 Google AppEngine 应用代管服务；IBM 公司的虚拟资源池提供企业云计算整合方案；Oracle 公司的 E2 上的 Oracle 数据库、OracleVM 和 Sun xVM；Amazon 公司的 EC2、C3（Simple Storage Server，简单存储服务）、SimpleDB 和 SQS。

第 2 章　初识 OpenStack

　　OpenStack 是一个由 NASA（美国国家航空航天局）和 Rackspace 合作研发并发起的，以 Apache 许可证授权的自由软件和开放源代码项目，由多个主要的组件组合起来完成具体工作。OpenStack 也是一个旨在为公共及私有云的建设与管理提供软件的开源项目。

　　OpenStack 支持几乎所有类型的云环境，项目目标是提供实施简单、可大规模扩展、丰富、标准统一的云计算管理平台。OpenStack 通过各种互补的服务提供了基础设施即服务（IaaS）的解决方案，每个服务提供 API 以进行集成。

　　OpenStack 云计算平台，帮助服务商和企业内部实现类似于 Amazon EC2 和 S3 的云基础架构服务（IaaS）。

2.1　OpenStack 发展历程

　　OpenStack 项目最初包含两个主要模块：Nova 和 Swift，前者是 NASA 开发的虚拟服务器部署和业务计算模块；后者是 Rackspace 开发的分布式云存储（对象存储）模块。2010 年 10 月，用于镜像管理的部件 Glance 也加入其中，进而形成了 OpenStack 的核心架构。

　　OpenStack 项目由 Rackspace 负责管理，2011 年 Rackspace 联合部分成员成立了 OpenStack 基金会，作为一个独立的组织，其下有三个分支：技术委员会、用户委员会和董事会。

　　OpenStack 发展迅速，成员既包括 IT 厂商、公司，也包括以个人名义或代表公司加入的个人成员，同时还吸引了众多 IT 巨头的加入和支持，成立了仅次于 Linux 的世界第二大开源基金会。

　　目前有超过 200 多家厂商支持 OpenStack，包括 Cisco、DELL、HPE、IBM、Intel、Oracle、Rackspace、RedHat 以及 VMware 等，还有一个大型的开发者社区致力于这个开源项目。OpenStack 称 Newton 版本是由 309 个企业机构的 2581 名开发者开发的。

　　OpenStack 可以让用户通过一个仪表板控制数据中心内大型的计算、存储和网络资源池，并通过一个 Web 接口配置资源。OpenStack 瞄准的是那些希望构建公有云或者私有云服务的提供商、企业、政府机构以及学术机构。

2.2　OpenStack 概述

　　OpenStack 的愿景是为所有公有云和私有云提供商提供可满足其任意需求、容易实施且可以大规模扩展的开源云计算平台。整个 OpenStack 项目被设计为可大规模灵活扩展的云计算操作系统,任何组织均可以通过 OpenStack 基于标准化的硬件设施创建和提供云计算服务。

亚马逊经过多年发展并取得巨大的成功,已成为事实上的 IaaS 的标准。

OpenStack 希望通过标准化服务的方式,在云计算的标准化和规范化方面有所推动。由于 OpenStack 的定位是亚马逊 AWS(Amazon Web Services)的源实现,无论在功能上还是 API 接口上,都尽可能与 AWS 保持兼容,因此,OpenStack 很多功能与亚马逊基本上是对应的(参见表 2 – 1)。

表 2 – 1 OpenStack 与 AWS 对照

AWS	OpenStack
EC2 弹性虚拟机	Nova 虚拟批管理程序
S3 云存储	Swift 对象存储组件
EBS 弹性云硬盘	Nova-volume / Cinder 虚拟机的存储管理组件
ELB 负载均衡	Atlas-LB 实现负载均衡(OpenStack 外围项目)
Console 控制台	Dashboard Herizon 界面访问控制台
VPC 虚拟私有云	Neutron 网络管理的组件
IAM 认证鉴权	Keystone 提供身份认证和授权的组件
Elastic MapReduce	Sahara 大数据方案

OpenStack 提供了框架标准和 API,用户可以以此为基础构建云计算解决方案。OpenStack 所有模块子系统之间均是通过标准虚拟化的 API 实现服务调用的。OpenStack 支持融合诸多厂商的云计算服务,如 KVM、VMware、Xen、Hyper-V 等。

OpenStack 开源软件以 Apache 许可证授权。OpenStack 的版本由 OpenStack 基金会整理与发布。

OpenStack 主要用 Python 编写,其代码质量相当高,带有一个完全文档化的 API。由于源代码是公开的,因此,代码中的质量问题、安全漏洞更易被发现并被修正。

2.3　OpenStack 架构

OpenStack 具有如下功能:项目所有的构成子系统和服务均被集成起来,一起提供 IaaS 服务,通过标准化公用服务接口 API 实现集成,子系统和服务之间可以通过 API 互相调用。

OpenStack 采用了职责拆分的设计理念,根据职责不同拆分成 7 个核心系统,每个系统都可以独立部署和使用。在每个子系统中,又根据分层(layer)设计理念,拆分成

API、逻辑处理（包括数据库存储）和底层驱动适配 3 个层次。

OpenStack 核心系统概念架构，如图 2-1 所示。

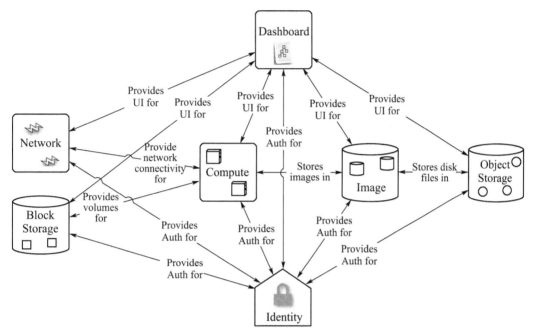

图 2-1　OpenStack 核心系统概念架构

用户可根据需要，选择将 OpenStack 各服务组件安装到一个或多个主机节点中。

如果仅仅用于一般性测试目的，则通常会采用 All in one 部署方式，即将 OpenStack 所有的服务组件安装到同一个主机节点中。

在生产环境下，都会采用多节点部署方式，各主机节点安装不同的服务组件承担不同的服务角色，平台易于扩展、伸缩性强、功能更丰富。

多节点部署 OpenStack 平台，则通常会采用以下两种典型架构：

1. 两节点架构

此架构中包含两种类型的节点，分别是控制节点（Control Node）和计算节点（Compute Node），控制节点只有一个，计算节点可以有多个。

控制节点运行所有控制服务，包括认证服务（Keystone）、镜像服务（Glance）、仪表板（Horizon）、网络组件（Neutron）和计算组件（Nova）的管理部分，还包含相关的 API 服务、MySQL 数据库和消息系统。

计算节点运行计算组件的虚拟机监控程序（Hypervisor），用来操作租户的虚拟机和实例。默认情况下，计算服务使用 KVM 作为管理程序，同时计算节点也运行网络插件和第二层代理，用来操作租户网络和实现安全组。计算节点可以运行多个。

两节点部署架构，如图 2-2 所示。

9

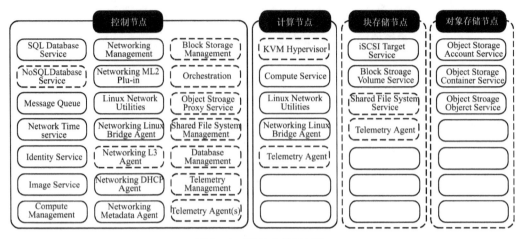

图2-2 两节点部署架构

2. 三节点架构

此架构中包含三种类型的节点，分别是控制节点（Control Node）、网络节点（Network Node）和计算节点（Compute Node），控制节点只有一个，计算节点可以有多个。

控制节点运行大部分控制服务，包括认证服务（Keystone）、镜像服务（Glance）、计算组件（Nova）和网络组件（Neutron）的管理部分、网络插件和仪表板（Horizon），还包含相关的API服务、MySQL数据库和消息系统。控制节点只有一个。

网络节点运行网络插件代理（Plugin Agent）、第二层代理（L2 Agent）和若干个第三层代理（L3 Agent），用来规划并操作租户（Tenant）的网络。第二层代理提供虚拟网络和隧道，第三层代理提供NAT和DHCP服务，也能处理租户虚拟机或实例的内外部连接。

计算节点运行计算组件的虚拟机监控程序（Hypervisor），用来操作租户的虚拟机和实例。默认情况下，计算服务使用KVM作为管理程序。同时计算节点也运行网络插件和第二层代理，用来操作租户网络和实现安全组。计算节点可以运行多个。

三节点部署架构，如图2-3所示。

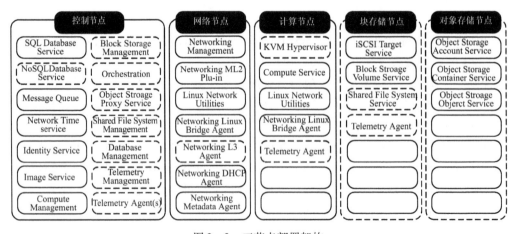

图2-3 三节点部署架构

2.4 OpenStack 服务组件

在建立 OpenStack 系统时，OpenStack 的核心服务组件是必须安装的，它们分别是 Keystone、Glance、Nova、Neutron。虽然其他服务组件可以进行选择性安装，但通常都会部署一个功能完整的 OpenStack 平台架构。

表 2-2 OpenStack 服务组件

服务名称	项目名称	功能描述
Dashboard 仪表板	Horizon	提供了一个基于 Web 的自服务门户，与 OpenStack 底层服务交互，诸如启动一个实例，分配 IP 地址以及配置访问控制
Compute 计算服务	Nova	在 OpenStack 环境中计算实例的生命周期管理。按需响应包括生成、调度、回收虚拟机等在内的操作
Networking 网络服务	Neutron	确保为其他 OpenStack 服务提供网络连接即服务，比如 OpenStack 计算。为用户提供 API 定义网络和使用。基于插件的架构其支持众多的网络提供商和技术
存储服务		
Block Storage 块存储服务	Cinder	为运行实例而提供的持久性块存储。它的可插拔驱动架构的功能有助于创建和管理块存储设备
Object Storage 对象存储服务	Swift	通过一个 RESTful，基于 HTTP 的应用程序接口存储和任意检索的非结构化数据对象。它拥有高容错机制，基于数据复制和可扩展架构。它的实现像是一个文件服务器需要挂载目录。在此种方式下，它写入对象和文件到多个硬盘中，以确保数据是在集群内跨服务器的多份复制
公共服务		
Identity service 认证服务	Keystone	为其他 OpenStack 服务提供认证和授权服务，为所有的 OpenStack 服务提供一个端点目录
Image service 镜像服务	Glance	存储和检索虚拟机磁盘镜像，OpenStack 计算会在实例部署时使用此服务
Telemetry 计量服务	Ceilometer	为 OpenStack 云的计费、基准、扩展性以及统计等目的提供监测和计量
高级服务		
Orchestration 编排服务	Heat	Orchestration 服务支持多样化的综合的云应用，通过调用 OpenStack-native REST API 和 CloudFormation-compatible Query API，支持 HOT 格式模板或者 AWS CloudFormation 格式模板，让用户能够通过模板来管理资源

整个 Openstack 对终端用户提供两大类访问入口：Horizon 和 API。所有子系统提供标准化 API，终端用户（包括开发人员）通过 API 访问和调用不同子系统的服务。子系统内部划分 API、逻辑处理（Manager）和底层驱动适配（Driver）。

不同子系统通过 API 实现交互，这里主要是指 REST（标准化接口）风格的 API。RESTfull 网格的接口设计理念基于 HTTP 协议，类似于 WebService，但更简单。

OpenStack 是由多个独立的部分组成的，这些独立的部分称之为 OpenStack 服务。所有服务的认证都通过一个通用的身份服务。分离的服务通过公开的 API 来进行彼此之间交互，除非那里需要管理员命令的特权。

OpenStack 系统服务的概念架构，如图 2-4 所示。

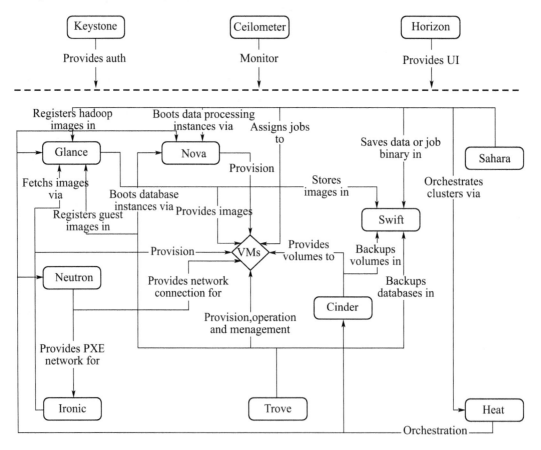

图 2-4　OpenStack 系统服务的概念架构

为了设计、部署和配置 OpenStack，管理员必须理解逻辑架构。

在 OpenStack 服务的内部往往由多个进程组成。所有的服务至少拥有一个 API 进程，用来监听 API 的请求，预先处理它们以及将它们传递给服务的其他部分。除了身份服务之外，其他服务的实际工作都不是由一个进程来完成的。

对于服务中的进程之间的通信，使用了 AMQP 消息管家。服务的状态存储在数据库中。当部署和配置 OpenStack 云时，可以从很多种消息管家和数据库解决方案中来做出选择，例如 RabbitMQ、Qpid、MySQL、MariaDB 以及 SQL。

用户可以通过基于通过 dashboard service 所实现的 Web 用户界面来访问 OpenStack，或者通过 via 命令行客户端，又或者是通过一些诸如浏览器插件或 curl 之类的工具来请求 API。对于应用来说，可使用 several SDKs。最终，所有的这些访问方法都是调用各个 OpenStack 服务的 REST API。

OpenStack 系统服务逻辑架构，如图 2-5 所示。

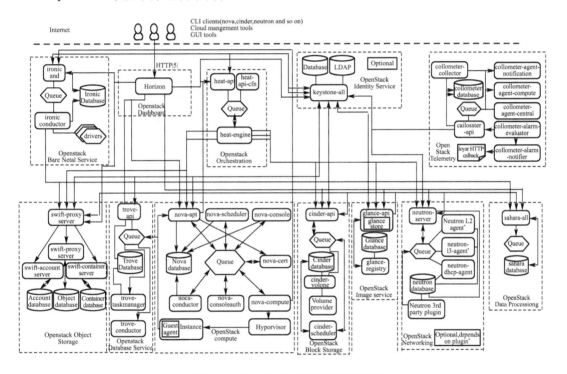

图 2-5　OpenStack 系统服务逻辑架构

2.5　OpenStack 版本演变

OpenStack 的技术更新和版本发生速度很快，从机构成立至第一个版本发布仅用了很短的时间，之后基本上每六个月发布一个新版本。

2010年6月，Rackspace和NASA成立OpenStack。同年7月，有超过25名合作伙伴，10月，发布了首个版本，代号为"Austin"，合作伙伴达到35名。

Open Stack的版本发展情况具体如表2-3所示。

表2-3 OpenStack版本发展

版本代号	生命周期开始时间/结束时间	功能演变
Queens	未开发	
Pike	未开发	
Ocata	2017-02-22/未定	
Newton	2016-10-06/未定	对Ironic裸机配置服务、Magnum容器调度集群管理器以及Kuryr容器网络的升级，将让集成容器、虚拟和物理基础架构到单一控制面板下变得更容易
Mitaka	2016-04-07/2017-04-10	重点在用户体验上简化了Nova、Keynote的使用，以及使用一致的API调用创建资源
Liberty	2015-10-15/2016-11-17	更面向企业，包括开始对跨一系列产品进行滚动升级的支持，以及对管理性和可扩展性的增强。引入了Magum容器管理，支持Kubernetes、Mesos和Docker Swarm
Kilo	2015-04-30/2016-05-02	裸机服务Ironic完全发布，增加互操作性
Juno	2014-10-16/2015-12-07	对现有功能的更进一步增强，推出数据服务，Trove和Sahara。其他重要的改变还包括对滚动升级的改进，网络，以及对Docker和裸机的支持
Icehouse	2014-04-17/2015-07-02	增加了Trove数据库服务，第一次滚动升级计算节点和Keynote联合认证，自动扩展和计量的功能
Havana	2013-10-17/2014-09-30	正式发布Ceilometer项目和Heat项目，Quantum网络服务变更为Neutron
Grizzly	2013-04-04/2014-03-29	将Melang和Quantum融合起来支撑网络服务
Folsom	2012-09-27/2013-11-19	增加了Cinder块存储，以及Quantum网络服务
Essex	2012-04-05/2013-05-06	发布了完整功能的Horizon仪表板服务和Keystone身份认证服务
Diablo	2011-09-22/2013-05-06	相对比较稳定的版本，可以开始大规模部署应用
Cactus	2011-04-15/不再支持	

续表

版本代号	生命周期开始时间/结束时间	功能演变
Bexar	2011-02-03/不再支持	
Austin	2010-10-21/不再支持	Rackspace 和 NASA 合作发布的第一个版本，结合了 Swift 存储和 Nova 计算模块

备注：以上数据截至 2017 年 3 月。

第3章 OpenStack 架构示例

3.1 部署架构

在本书中，将采用两节点架构方案作为示例架构进行 OpenStack 平台的部署工作。

如图 3-1 中所示，实线框表示必须安装的主机节点或服务组件。虚线框表示可选安装（可不安装）的主机节点或服务组件。

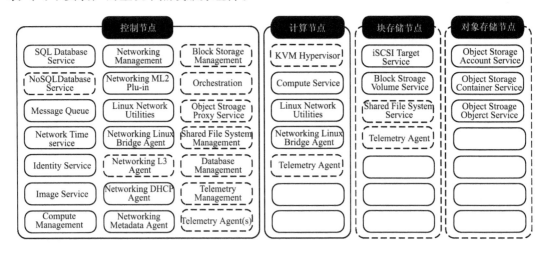

图 3-1 两节点架构方案

3.2 硬件配置

在本书示例中，将采用 5 台主机服务器，分别是控制节点（1 台）、计算节点（1 台）、块存储节点（1 台）和对象存储节点（2 台）。在此基础上可通过增加除控制节点之外的其他节点服务器，来扩展 OpenStack 平台架构。

为配合本书中 OpenStack 示例环境的部署需求，在这里提供了一个主机服务器的最小硬件配置。如果条件允许，可自行增加主机服务器的硬件配置。

各节点服务器的 CPU、内存、网卡、硬盘的硬件配置信息，如图 3-2 所示。

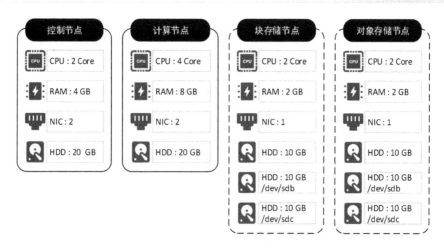

图 3-2 节点配置信息

3.3 服务组件

1. 控制节点（Controller Node）

控制节点上运行身份认证服务（Keystone）、镜像服务（Glance）、计算服务（Nova）的管理部分、网络服务（Neutron）的管理部分、多种网络代理（Proxy Services）以及仪表板（Horizon），同时也需要包含一些支持服务，例如：MySQL 数据库、NoSQL 数据库、消息队列（RabbitMQ）和 NTP。

控制节点上还可运行（可选的）部分块存储（Cinder）、对象存储（Swift）、编排服务（Heat）和计量服务（Ceilometer）。

控制节点上需要至少两块网卡。

2. 计算节点（Compute Node）

计算节点上运行计算服务（Nova）中管理实例的管理程序部分。默认情况下，计算服务使用 KVM，在本书中将配置使用 QEMU。

每个计算节点需要至少两块网卡。可以部署超过一个计算节点。

3. 块存储节点（Block Storage Node）

块存储节点上可运行（可选的）块存储服务（Cinder），共享文件服务（Manila）。块存储节点可以向实例提供这些存储磁盘。

为了简单起见，计算节点和块存储节点之间的服务流量使用管理网络。生产环境中应该部署一个单独的存储网络以增强性能和安全性。

每个块存储节点要求至少一块网卡。可以部署超过一个块存储节点。

4. 对象存储节点（Object Storage Node）

对象存储节点上可运行（可选的）对象存储服务（Swift）。对象存储服务用这些存储磁盘来存储账号、容器和对象。

为了简单起见，计算节点和对象存储节点之间的服务流量使用管理网络。生产环境中应该部署一个单独的存储网络以增强性能和安全性。

对象存储服务要求至少有两个节点，每个节点要求最少一块网卡。可以部署超过两个对象存储节点。

3.4 虚拟网络类型

OpenStack 平台中的网络分为两类，分别是公共网络（Provider networks）和私有网络（Self-service networks）。

3.4.1 公共网络（Provider networks）

公共网络（Provider networks）允许具有管理员权限的用户创建虚拟网络直接连接到物理网络中，实例利用此网络可以直接访问外部网络。其主要通过 Layer-2（网桥/交换机）服务以及划分 VLAN 网络来部署 OpenStack 网络服务。

公共网络（Provider networks）实质上建立的是虚拟网络到物理网络的桥，它依靠物理网络基础设施提供 Layer-3（路由）服务，额外的 DHCP 服务可为实例提供 IP 地址信息。其支持 Flat（untagged）和 VLAN（802.1Q tagged）两种网络类型。

3.4.2 私有网络（Self-service networks）

私有网络（Self-service networks）允许普通用户在租户内创建自有虚拟网络。在默认情况下，租户之间的网络是隔离的、不能共享的。

私有网络（Self-service networks）完全兼容公共网络（Provider networks）功能并进行了扩展，增加支持 Layer-3（路由）服务和 VXLAN 重叠网络技术。

私有网络（Self-service networks）实质上是使用 NAT 路由虚拟网络连接到物理网络，另外也为更高级的服务提供了基础，例如 LBaas 和 FWaaS。

3.5 主机网络

出于管理目的，例如：安装包、安全更新、DNS 和 NTP，所有的节点都需要被允许访问互联网。在大部分情况下，节点应该通过管理网络接口访问互联网。为了更好地突出网络隔离的重要性，示例架构中为管理网络使用私有 IP 地址（例如 192.168.1.X），并假定物理网络设备通过 NAT 或者其他方式提供互联网访问。示例架构使用可路由的 IP 地址隔离服务商（外部）网络并且假定物理网络设备直接提供互联网访问。

在公共网络架构中，所有实例直接连接到公共网络。在私有网络架构中，实例可以连接到自服务或公共网络。私有网络可以完全在 OpenStack 环境中或者通过外部网络使用 NAT 提供某种级别的外部网络访问,如图 3-3 所示。

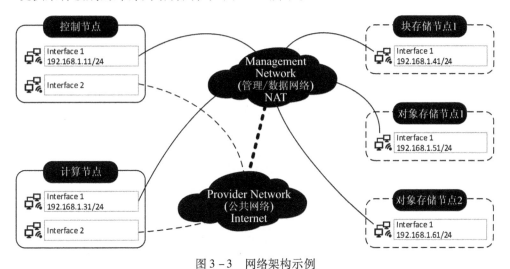

图 3-3 网络架构示例

简单来说,就是在进行 OpenStack 平台部署前,需要保证所有节点可以访问互联网,以进行系统安全更新和在线安装服务组件软件包。

3.6 密码配置

OpenStack 的各服务组件支持多种安全方式,如密码、策略和加密等,有的还支持数据库加密和消息代理。虽各有不同,但都支持设置密码的安全方式(参见表 3-1)。

本书为了简化 OpenStack 安装过程,对所有的服务组件约定一个密码配置。在实际生产环境中,密码设置需要确保其安全性,可以根据需求,替换这些约定的密码配置。

还可以使用命令生成复杂字符串,作为密码或 Token(令牌)使用,如生成一个随机的字符串(例如,用此值替换本书中的 ADMIN_TOKEN 值),以下是命令,接着图 3-4 是执行后信息。

openssl rand-hex 10

```
[root@localhost 桌面]# openssl rand -hex 10
5f5cb896a26399254da7
```

图 3-4 执行信息

生成一个随机的字符串哈希值(例如,用此值替换本书中的 swift_hash_path_suffix 和 swift_hash_path_prefix 值)的命令如下,结果如图 3-5 所示。

openssl rand-hex 10 | md5sum

```
[root@localhost 桌面]# openssl rand -hex 10 | md5sum
53260bc8e2efd4c0ed626ad18cab6795  -
```

图 3-5　执行结果

表 3-1　OpenStack 服务组件密码配置

密　码	用　途
SQL_PASS	SQL 数据库超级管理员用户（root）密码
RABBIT_PASS	消息队列服务（RabbitMQ）默认用户（guest）的密码
KEYSTONE_DBPASS	认证服务（Keystone）的数据库密码
ADMIN_TOKEN	认证服务（Keystone）初始配置时的管理员令牌（Token）
ADMIN_PASS	认证服务（Keystone）的超级管理员权限用户（admin）的密码
DEMO_PASS	认证服务（Keystone）的普通权限用户（demo）的密码
GLANCE_DBPASS	镜像服务（Glance）的数据库密码
GLANCE_PASS	镜像服务（Glance）用户（glance）的密码
NOVA_DBPASS	计算服务（Nova）的数据库密码
NOVA_PASS	计算服务（Nova）用户（nova）的密码
NEUTRON_DBPASS	网络服务（Neutron）的数据库密码
NEUTRON_PASS	网络服务（Neutron）用户（neutron）的密码
METADATA_SECRET	网络服务（Neutron）元数据代理（Metadata）的共享密码
CINDER_DBPASS	块存储服务（Cinder）的数据库密码
CINDER_PASS	块存储服务（Cinder）用户（cinder）的密码
MANILA_DBPASS	文件共享服务（Manila）的数据库密码
MANILA_PASS	文件共享服务（Manila）用户（manila）的密码
SWIFT_PASS	对象存储服务（Swift）用户（swift）的密码
HEAT_DBPASS	编排服务（Heat）的数据库密码
HEAT_PASS	编排服务（Heat）用户（heat）的密码
HEAT_DOMAIN_PASS	编排服务（Heat）域的密码
CEILOMETER_DBPASS	计量服务（Ceilometer）的数据库密码
CEILOMETER_PASS	计量服务（Ceilometer）用户（ceilometer）的密码
AODH_DBPASS	警告服务（Aodh）的数据库密码
AODH_PASS	警告服务（Aodh）用户（aodh）的密码

第 4 章 准备 OpenStack 初始环境

4.1 操作系统

操作系统建议使用最小化安装的 Linux 系统，以避免可能的软件包版本冲突以及为 OpenStack 提供更多资源。同时，需在每个节点安装 64 位发行版的 Linux 系统。

本书示例环境将使用 OpenStack Mitaka 版本来进行 OpenStack 的安装部署。在安装部署的过程中，需要运行大量 CLI 命令以及修改众多的服务组件配置文件。

桌面版的 CentOS 系统提供了 gedit 图形化的文本编辑程序，对于初学者而言，它更容易操作，如图 4-1 所示。

图 4-1 gedit 文本编辑程序

如果熟悉使用终端界面下的 vi 或 vim 工具，那么建议使用最小化安装的 CentOS 7.2 系统，如图 4-2 所示。

图 4-2 VIM-VI 工具

本书中使用 gedit 命令作为修改配置文件的默认工具程序，如果习惯使用 vi 或 vim 命令，那么请用 vi 或 vim 命令替换本书中的 gedit 命令。

在图形桌面以及文字终端环境下，OpenStack 的安装配置过程都是相同的。

4.2 进入终端

OpenStack 的安装配置过程需要在终端命令行下进行操作。在桌面环境下需要启动终端程序或切换到文字终端模式。若在文字终端环境下，则无须进行这一操作。

1. 启动终端程序

打开菜单"应用程序" > "收藏/工具" > "终端"，如图 4-3 所示。

图 4-3 启动终端程序

在弹出的终端程序窗口中,可以执行命令行操作,如图 4-4 所示。

图 4-4　命令行操作

2. 切换终端模式

通过快捷键,可在 GNOME 桌面环境和文字终端环境之间进行相互切换。

在 GNOME 桌面中,按 Ctrl + Alt + F2 组合键,可切换到文字终端界面,如图 4-5 所示。

图 4-5　文字终端界面

在文字终端界面中,按 Ctrl + Alt + F1 组合键,可切换到 GNOME 桌面,如图 4-4 所示。

4.3 关闭防火墙（所有节点）

为保持 OpenStack 服务组件的网络连通性，需要关闭 CentOS 系统防火墙。

4.3.1 关闭 selinux

1）关闭 selinux
setenforce 0
2）禁用 selinux
编辑 /etc/sysconfig/selinux 文件：
gedit /etc/sysconfig/selinux
修改以下内容：

SELINUX = disabled

4.3.2 关闭 firewall

1）关闭 firewall
systemctl stop firewalld.service
2）禁用 firewall
systemctl disable firewalld

4.4 配置主机名（所有节点）

为 OpenStack 各节点主机配置相对应的主机名，如表 4-1 所示。

表 4-1　OpenStack 各节点主机名配置

主机节点	主机名
控制节点	Controller
计算节点 1	Compute01
块存储节点 1	BlockStorage01
对象存储节点 1	ObjectStorage01
对象存储节点 2	ObjectStorage02

4.4.1 配置主机名

使用 hostnamectl 命令配置主机名。

1）控制节点

\# hostnamectl--static set-hostname controller

2）计算节点1

\# hostnamectl--static set-hostname compute01

3）块存储节点1

\# hostnamectl--static set-hostname blockstorage01

4）对象存储节点1

\# hostnamectl--static set-hostname objectstorage01

5）对象存储节点2

\# hostnamectl--static set-hostname objectstorage02

4.4.2 验证操作

使用 hostname 命令查看并验证主机名配置的正确性。

\# hostname

4.5 配置网络（所有节点）

为 OpenStack 各节点主机配置相对应的网络，如图 4-6 所示。

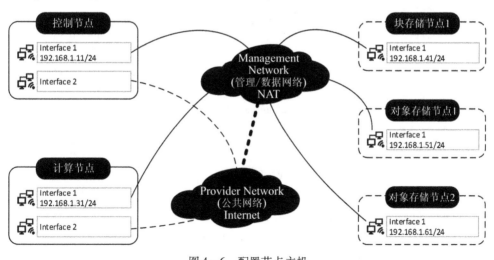

图 4-6 配置节点主机

25

在本书的示例架构中，各节点主机中的第 1 块网卡将用于管理/数据网络（私有网络）。控制节点和计算节点的第 2 块网卡用于公共网络，它可连接到物理网络，以提供互联网访问。

CentOS 7.2 系统的网卡接口命名采用了可预见命名规则。这一规则，使接口名称被自动基于固件、拓扑结构和位置信息来确定。

在本书的示例架构中，约定第 1 块网卡的设备名称是 eno16777984，第 2 块网卡的设备名称是 eno33557248。

可以使用 ip addr 命令，查看网卡设备信息，如图 4-7 所示。

ip addr

```
[root@localhost 桌面]# ip addr
1: lo: <LOOPBACK,UP,LOWER_UP> mtu 65536 qdisc noqueue state UNKNOWN
    link/loopback 00:00:00:00:00:00 brd 00:00:00:00:00:00
    inet 127.0.0.1/8 scope host lo
       valid_lft forever preferred_lft forever
    inet6 ::1/128 scope host
       valid_lft forever preferred_lft forever
2: eno16777984: <BROADCAST,MULTICAST,UP,LOWER_UP> mtu 1500 qdisc mq state UP qlen 1000
    link/ether 00:50:56:a8:8f:52 brd ff:ff:ff:ff:ff:ff
3: eno33557248: <BROADCAST,MULTICAST,UP,LOWER_UP> mtu 1500 qdisc mq state UP qlen 1000
    link/ether 00:50:56:a8:e4:f3 brd ff:ff:ff:ff:ff:ff
```

图 4-7　网卡设备信息

CentOS 7.2 系统的网卡接口配置文件保存在 /etc/sysconfig/network-scripts 目录中，接口配置文件名格式是 ifcfg-INTERFACE_NAME。

可以使用 ls 命令，查看网卡接口配置文件，如图 4-8 所示。

```
[root@localhost 桌面]# ls /etc/sysconfig/network-scripts/
ifcfg-eno16777984   ifdown-ipv6    ifdown-tunnel   ifup-isdn     ifup-TeamPort
ifcfg-eno33557248   ifdown-isdn    ifup            ifup-plip     ifup-tunnel
ifcfg-lo            ifdown-post    ifup-aliases    ifup-plusb    ifup-wireless
ifdown              ifdown-ppp     ifup-bnep       ifup-post     init.ipv6-global
ifdown-bnep         ifdown-routes  ifup-eth        ifup-ppp      network-functions
ifdown-eth          ifdown-sit     ifup-ib         ifup-routes   network-functions-ipv6
ifdown-ib           ifdown-Team    ifup-ippp       ifup-sit
ifdown-ippp         ifdown-TeamPort ifup-ipv6      ifup-Team
```

图 8-4　网卡接口配置文件

可见，网卡设备名称与其接口配置文件是一一对应的。第 1 块网卡的设备名称是 eno16777984，与其对应的接口配置文件是 ifcfg-eno16777984，第 2 块网卡的设备名称是 eno33557248，与其对应的接口配置文件是 ifcfg-eno33557248。

如果在 /etc/sysconfig/network-scripts 目录中找不到与网卡设备名称对应的接口配置文件，则可以手动创建这个网卡接口配置文件。

下面是一个网卡接口配置文件的默认内容示例：

```
TYPE = Ethernet
BOOTPROTO = dhcp
DEFROUTE = yes
IPV4_FAILURE_FATAL = no
IPV6INIT = yes
IPV6_AUTOCONF = yes
IPV6_DEFROUTE = yes
IPV6_FAILURE_FATAL = no
NAME = eno16777984
UUID = 7e6d9147-f075-4d3e-b0bf-5f0c0d117180
DEVICE = eno16777984
ONBOOT = yes
PEERDNS = yes
PEERROUTES = yes
IPV6_PEERDNS = yes
IPV6_PEERROUTES = yes
```

在本书的示例架构中，仅需要为第 1 块网卡配置管理网络，第 2 块网卡无须进行配置。

表 4-2 是各节点主机管理网络配置。

表 4-2 OpenStack 各节点主机管理网络配置

主机节点	Management Network（管理/数据网络）	
	（Interface 1，eno16777984）	
控制节点	192.168.1.11	255.255.255.0
计算节点 1	192.168.1.31	255.255.255.0
块存储节点 1	192.168.1.41	255.255.255.0
对象存储节点 1	192.168.1.51	255.255.255.0
对象存储节点 2	192.168.1.61	255.255.255.0

4.5.1 配置管理网络

1. 控制节点

1）修改第 1 块网卡设备（eno16777984）的接口配置文件

gedit /etc/sysconfig/network-scripts/ifcfg-eno16777984
修改或添加以下内容：

TYPE = Ethernet
BOOTPROTO = none
ONBOOT = yes
IPADDR = 192.168.1.11
NETMASK = 255.255.255.0
GATEWAY = 192.168.1.254

2）重新启动网络服务
systemctl restart network.service

2. 计算节点1

1）修改第1块网卡设备（eno16777984）的接口配置文件
gedit /etc/sysconfig/network-scripts/ifcfg-eno16777984
修改或添加以下内容：

TYPE = Ethernet
BOOTPROTO = none
ONBOOT = yes
IPADDR = 192.168.1.31
NETMASK = 255.255.255.0
GATEWAY = 192.168.1.254

2）重新启动网络服务
systemctl restart network.service

3. 块存储节点1

1）修改第1块网卡设备（eno16777984）的接口配置文件
gedit /etc/sysconfig/network-scripts/ifcfg-eno16777984
修改或添加以下内容：

TYPE = Ethernet
BOOTPROTO = none
ONBOOT = yes
IPADDR = 192.168.1.41
NETMASK = 255.255.255.0
GATEWAY = 192.168.1.254

2）重新启动网络服务

systemctl restart network.service

4. 对象存储节点1

1）修改第1块网卡设备（eno16777984）的接口配置文件

gedit /etc/sysconfig/network-scripts/ifcfg-eno16777984

修改或添加以下内容：

```
TYPE = Ethernet
BOOTPROTO = none
ONBOOT = yes
IPADDR = 192.168.1.51
NETMASK = 255.255.255.0
GATEWAY = 192.168.1.254
```

2）重新启动网络服务

systemctl restart network.service

5. 对象存储节点2

1）修改第1块网卡设备（eno16777984）的接口配置文件

gedit /etc/sysconfig/network-scripts/ifcfg-eno16777984

修改或添加以下内容：

```
TYPE = Ethernet
BOOTPROTO = none
ONBOOT = yes
IPADDR = 192.168.1.61
NETMASK = 255.255.255.0
GATEWAY = 192.168.1.254
```

2）重新启动网络服务

systemctl restart network.service

4.5.2 验证操作

在所有节点主机中，相互测试到其他节点的IP连通性，如下所示。

ping-c 4 192.168.1.11

ping-c 4 192.168.1.31

ping-c 4 192.168.1.41

ping-c 4 192.168.1.51

ping-c 4 192.168.1.61

4.6 配置 DNS 解析（所有节点）

在本书示例架构中，没有部署 DNS 服务器，需要修改各节点主机的 hosts 文件，让各节点主机之间可以正确解析主机名。

4.6.1 配置 hosts 文件

1）编辑 /etc/hosts 文件
gedit /etc/hosts
在文件尾部添加以下内容：

```
192.168.1.11  controller
192.168.1.31  compute01
192.168.1.41  blockstorage01
192.168.1.51  objectstorage01
192.168.1.61  objectstorage02
```

4.6.2 验证操作

1）在各节点主机中，相互测试主机名的连通性
ping-c 4 controller
ping-c 4 compute01
ping-c 4 blockstorage01
ping-c 4 objectstorage01
ping-c 4 objectstorage02

4.7 配置本地安装源（所有节点）

如果各节点主机可以访问互联网，并且希望使用在线安装源来安装 OpenStack 服务组件，那么请略过此小节的操作。

请务必在所有节点主机中完成本节的操作。

当节点主机没有条件访问互联网时，还可以使用提前下载好的服务组件 rpm 安装包进行离线安装。

config 目录：提供了 OpenStack 各服务组件的配置文件范例。

data 目录：提供了 OpenStack 安装配置所需的数据文件，如图 4-9 所示。

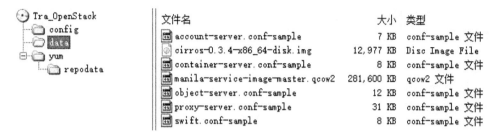

图 4-9 data 目录数据文件

yum 目录：提供了 OpenStack 各服务组件所需的 rpm 安装包文件，如图 4-10 所示。

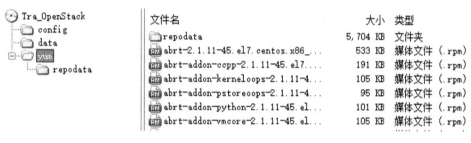

图 4-10 yum 目录安装包文件

4.7.1 配置本地安装源

将光盘中的数据文件复制到主机的本地磁盘，并配置成为本地安装源。
1) 挂载光盘驱动器
mount /dev/cdrom /mnt
2) 创建 OpenStack 目录，后续安装部署操作会从此目录中读取所需文件
mkdir /openstack
3) 从光盘中复制文件
cp-rf /mnt/ * /openstack
4) 卸载光盘驱动器
umount /mnt
5) 删除系统默认的安装源配置文件
rm-rf /etc/yum.repos.d/ *
6) 创建本地 yum 源配置文件
gedit /etc/yum.repos.d/openstack.repo
添加以下内容：

[openstack]
name = openstack repo
baseurl = file:///openstack/yum/
enabled = 1
gpgcheck = 0

7）清除 yum 缓存数据（见图 4-11）

yum clean all

```
[root@controller 桌面]# yum clean all
已加载插件：fastestmirror, langpacks
正在清理软件源： openstack
Cleaning up everything
```

图 4-11　清除 yum 缓存

如果遇到类似于"/var/run/yum.pid 已被锁定，PID 为 3148 的另一个程序正在运行。"的错误提示时，请先执行下面的命令，然后再重新执行上面的命令，结果如图 4-12 所示。

rm -f /var/run/yum.pid

```
[root@localhost 桌面]# yum clean all
已加载插件：fastestmirror, langpacks
/var/run/yum.pid 已被锁定，PID 为 3845 的另一个程序正在运行。
Another app is currently holding the yum lock; waiting for it to exit...
另一个应用程序是：PackageKit
    内存：142 M RSS (1.4 GB VSZ)
    已启动： Fri Mar 24 17:06:19 2017 - 03:28之前
    状态    ：睡眠中，进程ID：3845
```

图 4-12　执行清除缓存命令的结果

4.7.2　验证操作

验证 repo 源以及可安装的软件包，是否正确指向了自定义的 openstack 本地源。

1）查看 repo 源列表（结果见图 4-13）

yum repolist all

```
[root@controller 桌面]# yum repolist all
已加载插件：fastestmirror, langpacks
openstack                                          | 3.0 kB  00:00:00
openstack/primary_db                               | 1.7 MB  00:00:00
Determining fastest mirrors
源标识                          源名称                          状态
openstack                       openstack repo                  启用: 1,247
repolist: 1,247
```

图 4-13　查看 repo 源列表结果

2）查看软件包列表（结果见图4-14）

yum list

```
[root@localhost 桌面]# yum list
已加载插件：fastestmirror, langpacks
Loading mirror speeds from cached hostfile
已安装的软件包
GConf2.x86_64                          3.2.6-8.el7                      @anaconda
LibRaw.x86_64                          0.14.8-5.el7.20120830git98d925   @anaconda
ModemManager.x86_64                    1.1.0-8.git20130913.el7          @anaconda
ModemManager-glib.x86_64               1.1.0-8.git20130913.el7          @anaconda
可安装的软件包
GeoIP.x86_64                           1.5.0-11.el7                     openstack
ModemManager.x86_64                    1.6.0-2.el7                      openstack
ModemManager-glib.x86_64               1.6.0-2.el7                      openstack
MySQL-python.x86_64                    1.2.5-1.el7                      openstack
NetworkManager.x86_64                  1:1.4.0-13.el7_3                 openstack
NetworkManager-adsl.x86_64             1:1.4.0-13.el7_3                 openstack
NetworkManager-bluetooth.x86_64        1:1.4.0-13.el7_3                 openstack
NetworkManager-glib.x86_64             1:1.4.0-13.el7_3                 openstack
```

图4-14 查看软件包列表结果

4.8 配置NTP（所有节点）

本书示例中使用Chrony程序提供NTP（网络时间协议）服务，以同步各节点的系统时间。在控制器节点引用外部NTP服务器或使用系统主机时间，其他节点引用控制节点。

4.8.1 安装和配置

1. 控制节点

1）安装chrony软件包

yum -y install chrony

2）编辑 /etc/chrony.conf 文件

gedit /etc/chrony.conf

添加以下内容：

allow 192.168.1.0/24

如果控制节点无法访问互联网，可优先使用系统本地时钟作为时间同步源，添加以下内容：

server controller iburst
local stratum 10

3) 启动、重启 NTP 服务,并将其配置为随系统启动

\# systemctl enable chronyd.service

\# systemctl start chronyd.service

\# systemctl restart chronyd.service

2. 其他节点

1) 安装 chrony 软件包

\# yum-y install chrony

2) 编辑 /etc/chrony.conf 文件

\# gedit /etc/chrony.conf

在配置文件中,注释以下内容:

\# server 0.centos.pool.ntp.org iburst
\# server 1.centos.pool.ntp.org iburst
\# server 2.centos.pool.ntp.org iburst
\# server 3.centos.pool.ntp.org iburst

并添加以下内容:

server controller iburst

3) 启动、重启 NTP 服务,并将其配置为随系统启动

\# systemctl enable chronyd.service

\# systemctl start chronyd.service

\# systemctl restart chronyd.service

4.8.2 验证操作

1. 控制节点

验证 NTP 服务:

\# chronyc sources

当控制节点与 NTP 服务器同步成功时,将返回时间同步信息。S 列显示为 * 号,表示当前同步成功的 NTP 服务器,如图 5-15 所示。

```
[root@localhost 桌面]# chronyc sources
210 Number of sources = 4
MS Name/IP address         Stratum Poll Reach LastRx Last sample
===============================================================================
^- 161.53.131.133                2    6    37     17   +37ms[  +37ms] +/-  174ms
^+ ntp2.flashdance.cx            2    6    37     12   +18ms[  +30ms] +/-  156ms
^* 85.199.214.100                1    6    37      2   -30ms[  -18ms] +/-  119ms
^? marla.ludost.net              0    7     0    10y    +0ns[   +0ns] +/-    0ns
```

图 5-15 返回时间同步信息

当控制节点与 NTP 服务器无法同步时，则不会返回时间同步结果，如图 5-16 所示。

```
[root@controller 桌面]# chronyc sources
210 Number of sources = 0
MS Name/IP address         Stratum Poll Reach LastRx Last sample
===============================================================================
```

图 5-16　不同步结果

2. 其他节点

验证 NTP 服务：

chronyc sources

其他节点将返回与控制节点的时间同步结果，如图 5-17 所示。

```
[root@localhost 桌面]# chronyc sources
210 Number of sources = 1
MS Name/IP address         Stratum Poll Reach LastRx Last sample
===============================================================================
^? controller                    0    6     0    10y   +0ns[  +0ns] +/-    0ns
```

图 5-17　同步结果

4.9　启用 OpenStack 库（所有节点）

1）安装用于启用 OpenStack 仓库的软件包

yum -y install centos-release-openstack-mitaka

如果正在使用在线安装源，那么请略过下面的操作步骤。

安装后，会自动生成在线安装源配置文件，它会与原本配置的本地安装源相冲突，需要删除掉这些以 CentOS 开头的在线源配置文件。

2）删除在线安装源配置文件

rm -rf /etc/yum.repos.d/CentOS*

3）清除 yum 缓存数据

yum clean all

4.10　更新操作系统（所有节点）

在开始进行 OpenStack 安装部署之前，需要将操作系统更新至最新版本。

1）更新系统

yum -y upgrade

如果正在使用在线安装源，那么请略过下面的操作步骤。

更新完成后，会自动生成在线安装源配置文件，它会与原本配置的本地安装源相冲突，需要删除掉这些以 CentOS 开头的在线源配置文件。

2）删除在线安装源配置文件
rm-rf /etc/yum.repos.d/CentOS*
3）清除 yum 缓存数据
yum clean all
4）重启系统，以应用系统更新
reboot

4.11 安装 OpenStack 客户端（所有节点）

1）安装 OpenStack 客户端软件包
yum-y install python-openstackclient

CentOS 默认启用了 SELinux。安装 openstack-selinux 软件包以便自动管理 OpenStack 服务的安全策略。

2）安装 OpenStack selinux 软件包
yum-y install openstack-selinux

4.12 安装 SQL 数据库（控制节点）

大多数 OpenStack 服务使用 SQL 数据库来存储信息，SQL 数据库一般在控制节点上运行。

OpenStack 支持 MariaDB、MySQL、PostgreSQL 数据库，本书示例中使用的是 MariaDB 数据库，可根据实际需求，选择不同的数据库。

1）安装 MariaDB 软件包
yum-y install mariadb mariadb-server python2-PyMySQL
2）创建 /etc/my.cnf.d/openstack.cnf 文件
gedit /etc/my.cnf.d/openstack.cnf
添加以下内容：

```
[mysqld]
bind-address = SQL_BIND_IP_ADDRESS
default-storage-engine = innodb
innodb_file_per_table
max_connections = 4096
collation-server = utf8_general_ci
character-set-server = utf8
```

将 SQL_BIND_IP_ADDRESS 替换为控制节点的管理网络 IP 地址，本书示例中将其替换为 192.168.1.11，使得其他节点可以通过管理网络访问数据库，同时启用一组有用的选项和 UTF-8 字符集。

3）启动 MariaDB 服务，并将其配置为随系统启动

systemctl enable mariadb.service
systemctl start mariadb.service

4）运行 mysql_secure_installation 脚本，配置数据库服务的安全性

mysql_secure_installation

为数据库的 root 用户设置一个安全的密码以及设置安全选项。

本书示例中将 root 用户密码设置为 SQL_PASS，同时将安全选项设置为 y。

```
...
Set root password? [Y/n] y
New password:
Re-enter new password:
Password updated successfully!
Reloading privilege tables..
... Success!
...
Remove anonymous users? [Y/n] y
... Success!
...
Disallow root login remotely? [Y/n] y
... Success!
...
Remove test database and access to it? [Y/n] y
-Dropping test database...
... Success!
- Removing privileges on test database...
... Success!
...
Reload privilege tables now? [Y/n] y
... Success!
...
```

4.13 安装消息队列 RabbitMQ（控制节点）

OpenStack 使用 message queue 协调操作和服务之间的状态信息。消息队列服务一般在控制节点上运行。

OpenStack 支持好几种消息队列服务，包括 RabbitMQ、Qpid 和 ZeroMQ 等。不过，大多数发行版本的 OpenStack 软件包支持特定的消息队列服务。

本书示例中将安装 RabbitMQ 消息队列服务，因为大部分发行版本都支持它。

1）安装 RabbitMQ 软件包

\# yum -y install rabbitmq-server

2）启动消息队列服务，并将其配置为随系统启动

\# systemctl enable rabbitmq-server.service

\# systemctl start rabbitmq-server.service

3）创建 openstack 用户

\# rabbitmqctl add_user openstack RABBIT_PASS

本书示例中将 openstack 用户密码设置为 RABBIT_PASS，如图 5-18 所示。

```
[root@controller ~]# rabbitmqctl add_user openstack RABBIT_PASS
Creating user "openstack" ...
```

图 5-18 密码设置

4）设置 openstack 用户权限，使其可以配置、写和读 RabbitMQ（见图 5-19）

\# rabbitmqctl set_permissions openstack ".*" ".*" ".*"

```
[root@controller ~]# rabbitmqctl set_permissions openstack ".*" ".*" ".*"
Setting permissions for user "openstack" in vhost "/" ...
```

图 5-19 用户权限设置

4.14 安装缓存服务 Memcached（控制节点）

认证服务的认证缓存使用 Memcached 来缓存令牌（Token）。缓存服务 Memcached 运行在控制节点上。在生产环境中，推荐联合启用防火墙、认证和加密来保证它的安全。

1）安装 Memcached 软件包

\# yum -y install memcached python-memcached

2）启动 Memcached 服务，并将其配置为随系统启动

\# systemctl enable memcached.service

\# systemctl start memcached.service

第5章 身份认证（Keystone）服务

5.1 服务概述

Keystone 服务负责为 OpenStack 云平台提供通用的统一身份认证和管理服务（Identity Service）。它以 API 的方式为认证管理、授权管理和服务目录服务管理提供单点整合。

OpenStack 云平台中所有服务之间的鉴权和认证都需要经过它。通过 OpenStack 身份认证服务的认证之后，它会返回为一个在 OpenStack 各个服务之间传输用的鉴权密钥，接下来可以用这个密钥来为某个具体服务做鉴权和验证，并为用户创建合适的角色，以及服务、租户、用户帐户和 API Endpoints（端点）。

当某个 OpenStack 服务收到来自用户的请求时，该服务首先会询问身份认证服务，验证该用户是否有权限进行此次请求。

OpenStack 身份认证服务安装完成后，必须将 OpenStack 项目中的每个服务组件注册到其中，以使 OpenStack 身份认证服务组件能够识别这些服务组件并在网络中定位它们。

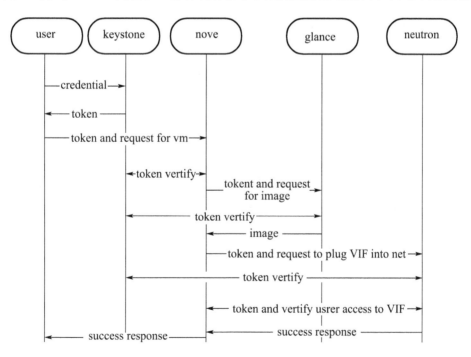

5.1.1 OpenStack 身份认证服务包含的组件

1. Server（服务器）

一个中心服务器（例如本书示例中的控制节点），使用 RESTful 程序接口来提供认证和授权服务。

2. Drivers（驱动）

驱动或服务后端被集成到中心服务器中，用来访问 OpenStack 外部项目和项目之内（例如 SQL 数据库）的程序访问身份认证信息。

3. Modules（模块）

中间件模块运行于使用身份认证服务的 OpenStack 组件的地址空间中。这些模块拦截服务请求，取出用户凭据，并将它们送入中心服务器寻求授权。中间件模块和 OpenStack 组件间的整合使用 Python Web 服务器网关接口。

5.1.2 OpenStack 身份认证服务基本概念

1. Service Entity（服务实体）

每个添加到 OpenStack 云平台的服务组件都需要在 Keystone 服务中创建一个服务实体（Service Entity）。

2. API Endpoint（API 端点）

一个可以通过网络访问的地址，使用 URL 和端口号的形式，用户可以通过一个使用 URL 和端口号形式的网络地址访问 OpenStack 服务组件。每个 Service Entity（服务实体）可以提供一个或多个 API Endpoint（API 端点）服务，用户可以通过它访问 OpenStack 资源并且可以执行相应的操作。

3. Tenant（租户）

Tenant（租户）在一些早期版本中被称为 Project（项目），它是各个服务中一些可以被访问的资源集合。例如，通过 Nova 创建虚拟机时要指定到某个租户中。用户访问租户的资源前，必须与该租户关联，并且指定在该租户下的角色。

一个租户可以有多个用户，一个用户可以同时属于多个不同的租户，当然在不同的租户中可以充当不同的角色，也即拥有不同的权限。

4. User（用户）

使用服务的用户，可以是人或者服务，抑或是系统中使用 OpenStack 相关服务的一个组织。

Keystone 身份认证服务组件会确认用户发出请求的有效性。用户可利用自身的 Token（令牌）登录和访问各种资源。

5. Role（角色）

定义一组用户的权利和权限，具备执行一系列指定操作的特性。Keystone 身份认证服务组件发送给用户的令牌环包含多个角色。当用户调用某个服务时，该服务会分析该用户的角色设置，确认该角色是否拥有操作和访问资源的权限。

OpenStack 中常用的角色有三种：admin、project manager、member。

admin 具有管理员权限

project manager 具有项目管理权限

member 具有项目使用权限

6. Credentials（证书）

用于确认用户身份的数据。例如：用户名和密码、用户名和 APIKey（键值）、Keystone 身份认证服务组件提供的认证令牌。

7. Authentication（认证）

对用户进行身份确认的一个过程。对于一个操作请求，Keystone 身份认证服务组件会验证发起请求的用户所提供的凭证。

凭证可以是用户名和密码、用户名和 API Key（键值）。当 Keystone 身份认证服务组件验证用户凭证正确后，会发出一个认证令牌。在后续的请求中，用户也会提供该认证令牌。

8. Token（令牌）

用于访问 OpenStack 项目 APIs 和由数字、字母组成的文本字符串。认证令牌有一个可用资源范围的限制，可以在任何时候被撤销，并且具有时效性。

Keystone 在对用户认证完毕后，为用户颁发一个 Token（令牌）。用户在随后的请求中，只需要亮出自己的令牌即可，而不需要发送自己的证书。

9. Domain（域）

项目和用户的集合，为认证实体定义了管理界限。它是个人、运营商或公司所拥有的空间，用户可以直接在域中进行管理操作。用户还可以获得域管理员角色，创建项目、用户和组，分配角色、用户和组。

10. Group（组）

组是域中拥有的所有用户的集合。组角色授予域或项目，应用于该组中所有的用户。从组中添加或删除用户会获得或撤销该用户在域或项目中相对应的角色和认证信息。

11. Region（区域）

一个有着专门用来与其他区域仅共享身份验证服务（Keystone）的专用 API 端点离散的 OpenStack 环境。

它可以关联多个子区域到同一个区域，形成树形结构的大区域。虽然区域并不代表实际的物理位置，但是可以使用代表物理区域的名词进行命名。例如：用 ch-gz 代表中国 –

广州。

12. OpenStackClient

OpenStack 项目中服务组件的命令行接口。例如，用户可以使用 openstack service create 和 nova boot 命令注册服务或创建实例。

5.2 安装前准备（控制节点）

在配置 OpenStack 身份认证服务前，必须创建一个数据库和管理员令牌。

5.2.1 创建数据库

1）用数据库连接客户端以 root 用户连接到数据库服务器
mysql-u root-p'SQL_PASS'

2）创建 keystone 数据库：
CREATE DATABASE keystone;

3）对 keystone 数据库授予恰当的权限
本书示例中将 keystone 数据库密码设置为 KEYSTONE_DBPASS。
GRANT ALL PRIVILEGES ON keystone. * TO 'keystone'@'localhost' \
  IDENTIFIED BY 'KEYSTONE_DBPASS';
GRANT ALL PRIVILEGES ON keystone. * TO 'keystone'@'%' \
  IDENTIFIED BY 'KEYSTONE_DBPASS';

4）退出数据库客户端
exit

5.3 安装和配置（控制节点）

5.3.1 安装软件包

使用带有 mod_wsgi 的 Apache HTTP 服务器来服务认证服务请求，端口为 5000 和 35357。缺省情况下，Kestone 服务会监听这些端口。

认证服务的认证缓存使用 Memcached 来缓存令牌（Token）。缓存服务 memecached 运行在控制节点上。在生产部署中，推荐联合启用防火墙、认证和加密来保证它的安全。

安装 keystone、httpd 软件包：
yum-y install openstack-keystone httpd mod_wsgi

5.3.2 配置 Keystone

1）编辑 /etc/keystone/keystone.conf 文件

由于默认配置文件在各发行版本中可能不同,因此,在进行修改的同时可能需要添加部分选项。另外在配置片段中的省略号(...)表示默认的配置选项,应该保留。

gedit /etc/keystone/keystone.conf

a. 在[DEFAULT]部分,定义初始管理令牌的值:

[DEFAULT]
...
admin_token = ADMIN_TOKEN

本书示例中将 admin_token 管理令牌的值设置为 ADMIN_TOKEN。

b. 在[database]部分,配置数据库访问:

[database]
...
connection = mysql+pymysql: //keystone: KEYSTONE_DBPASS@controller/keystone

c. 在[token]部分,配置 Fernet UUID 令牌的提供者:

[token]
...
provider = fernet

2)初始化身份认证服务的数据库

su-s /bin/sh-c "keystone-manage db_sync" keystone

3)初始化 Fernet keys

keystone-manage fernet_setup--keystone-user keystone \
--keystone-group keystone

5.3.3 配置 Apache

1)编辑 /etc/httpd/conf/httpd.conf 文件

gedit /etc/httpd/conf/httpd.conf

配置 ServerName 选项为控制节点主机名称:

ServerName controller

2)创建/etc/httpd/conf.d/wsgi-keystone.conf 文件

gedit /etc/httpd/conf.d/wsgi-keystone.conf

添加以下内容:

...
Listen 5000
Listen 35357

<VirtualHost *:5000>
    WSGIDaemonProcess keystone-public processes=5 threads=1 user=keystone group=keystone display-name=%{GROUP}
    WSGIProcessGroup keystone-public
    WSGIScriptAlias / /usr/bin/keystone-wsgi-public
    WSGIApplicationGroup %{GLOBAL}
    WSGIPassAuthorization On
    ErrorLogFormat "%{cu}t %M"
    ErrorLog /var/log/httpd/keystone-error.log
    CustomLog /var/log/httpd/keystone-access.log combined

    <Directory /usr/bin>
        Require all granted
    </Directory>
</VirtualHost>
<VirtualHost *:35357>
    WSGIDaemonProcess keystone-admin processes=5 threads=1 user=keystone group=keystone display-name=%{GROUP}
    WSGIProcessGroup keystone-admin
    WSGIScriptAlias / /usr/bin/keystone-wsgi-admin
    WSGIApplicationGroup %{GLOBAL}
    WSGIPassAuthorization On
    ErrorLogFormat "%{cu}t %M"
    ErrorLog /var/log/httpd/keystone-error.log
    CustomLog /var/log/httpd/keystone-access.log combined

    <Directory /usr/bin>
        Require all granted
    </Directory>
</VirtualHost>

5.3.4 完成安装

启动 Apache HTTP 服务，并将其配置为随系统启动：
systemctl enable httpd.service
systemctl start httpd.service

5.4 创建服务实体和 API 端点（控制节点）

Keystone 身份认证服务组件为 OpenStack 环境中其他服务组件的使用和操作提供了服务目录，每个添加到 OpenStack 环境中的服务组件在服务目录中需要一个服务实体（Service Entity）和一个 API 端点（API Endpoint）。

5.4.1 创建环境变量

默认情况下，身份认证服务数据库不包含支持传统认证和目录服务的信息。必须使用此前为身份认证服务创建的临时身份验证令牌用来初始化的服务实体和 API 端点。

为满足初始化认证服务的需要，使用环境变量配置临时认证令牌，以缩短命令行的长度。

1）配置认证令牌
export OS_TOKEN = ADMIN_TOKEN

2）配置端点 URL
export OS_URL = http：//controller：35357/v3

3）配置认证 API 版本
export OS_IDENTITY_API_VERSION = 3

5.4.2 创建服务实体和 API 端点

1. 创建服务实体和身份认证服务

在 Openstack 环境中认证服务管理服务目录。服务使用这个目录来决定环境中可用的服务。
openstack service create \
   --name keystone--description "OpenStack Identity" identity

运行结果如图 5-1 所示。

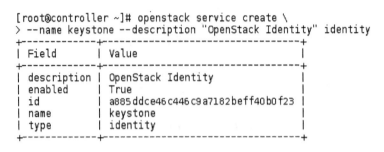

图 5-1 创建服务实体和身份认证服务

如果出现 HTTP 401 错误，如图 5-2 所示，则请先检查配置文件 /etc/keystone/keystone.conf 是否正确。例如：admin_token 项配置错误。然后重新初始化身份认证服务的数据库和 Fernet keys，并重启 Apache HTTP 服务。

```
[root@controller ~]# openstack service create \
> --name keystone --description "OpenStack Identity" identity
The request you have made requires authentication. (HTTP 401) (Request-ID: req-37800444-cb77-4171-89a5-af3a72cb9
4c8)
```

图 5-2 HTTP 401 错误

2. 创建 API 端点

身份认证服务管理了一个与环境相关的 API 端点目录。服务使用这个目录来决定如何与环境中的其他服务进行通信。

OpenStack 使用三个 API 端点变种代表每种服务：admin、internal 和 public。默认情况下，管理 API 端点允许修改用户和租户而公共和内部 APIs 不允许这些操作。在生产环境中，出于安全原因，变种是为了服务不同类型的用户能驻留在单独的网络上。对实例而言，公共 API 网络是为了让顾客在互联网上看见他们自己管理的云。管理 API 网络在管理云基础设施的组织中其操作也是有所限制的。内部 API 网络可能会被限制在包含 OpenStack 服务的主机上。此外，OpenStack 支持可伸缩性的多区域。为了简单起见，本书为所有端点变种和默认 RegionOne 区域都使用管理网络。

每个添加到 OpenStack 环境中的服务要求一个或多个服务实体和三个认证服务中的 API 端点变种。

openstack endpoint create--region RegionOne \
  identity public http：//controller：5000/v3

创建 public API 端点，结果信息如图 5-3 所示。

```
[root@controller ~]# openstack endpoint create --region RegionOne \
> identity public http://controller:5000/v3
+--------------+----------------------------------+
| Field        | Value                            |
+--------------+----------------------------------+
| enabled      | True                             |
| id           | 98304707440d4209ae3c07043895e2fc |
| interface    | public                           |
| region       | RegionOne                        |
| region_id    | RegionOne                        |
| service_id   | a885ddce46c446c9a7182beff40b0f23 |
| service_name | keystone                         |
| service_type | identity                         |
| url          | http://controller:5000/v3        |
+--------------+----------------------------------+
```

图 5-3 创建 public API 端点

openstack endpoint create--region RegionOne \
   identity internal http：//controller：5000/v3

创建 internal API 端点，结果信息如图 5-4 所示。

```
[root@controller ~]# openstack endpoint create --region RegionOne \
> identity internal http://controller:5000/v3
+--------------+----------------------------------+
| Field        | Value                            |
+--------------+----------------------------------+
| enabled      | True                             |
| id           | 84942d9e62ba4c8d82f9d95492308915 |
| interface    | internal                         |
| region       | RegionOne                        |
| region_id    | RegionOne                        |
| service_id   | a885ddce46c446c9a7182beff40b0f23 |
| service_name | keystone                         |
| service_type | identity                         |
| url          | http://controller:5000/v3        |
+--------------+----------------------------------+
```

图 5-4 创建 internal API 端点

openstack endpoint create--region RegionOne \
   identity admin http：//controller：35357/v3

创建 admin API 端点，结果信息如图 5-5 所示。

```
[root@controller ~]# openstack endpoint create --region RegionOne \
> identity admin http://controller:35357/v3
+--------------+----------------------------------+
| Field        | Value                            |
+--------------+----------------------------------+
| enabled      | True                             |
| id           | d8a1c01567e84c7a94e9dd5438dfc05c |
| interface    | admin                            |
| region       | RegionOne                        |
| region_id    | RegionOne                        |
| service_id   | a885ddce46c446c9a7182beff40b0f23 |
| service_name | keystone                         |
| service_type | identity                         |
| url          | http://controller:35357/v3       |
+--------------+----------------------------------+
```

图 5-5 创建 admin API 端点

5.5 创建域、项目、用户和角色（控制节点）

身份认证服务为每个 OpenStack 服务提供认证服务。认证服务使用域（domain）、项目（租户）、用户（user）和角色（role）的组合。

1）创建 Default Domain 域

openstack domain create--description "Default Domain" default

创建信息如图 5-6 所示。

```
[root@controller ~]# openstack domain create --description "Default Domain" default
+-------------+----------------------------------+
| Field       | Value                            |
+-------------+----------------------------------+
| description | Default Domain                   |
| enabled     | True                             |
| id          | 1224d01d12d5451191c2f0aeefa6ad6a |
| name        | default                          |
+-------------+----------------------------------+
```

图 5-6 创建 Default Domain 域

2）创建一个管理员权限的项目、用户和角色，用于管理任务操作

a. 创建 admin 项目：

openstack project create--domain default \

 --description "Admin Project" admin

如图 5-7 所示。

```
[root@controller ~]# openstack project create --domain default \
> --description "Admin Project" admin
+-------------+----------------------------------+
| Field       | Value                            |
+-------------+----------------------------------+
| description | Admin Project                    |
| domain_id   | 1224d01d12d5451191c2f0aeefa6ad6a |
| enabled     | True                             |
| id          | cec93d07f74a4346a3b2221241abb52e |
| is_domain   | False                            |
| name        | admin                            |
| parent_id   | 1224d01d12d5451191c2f0aeefa6ad6a |
+-------------+----------------------------------+
```

图 5-7　创建 admin 项目

b. 创建 admin 用户：

\# openstack user create--domain default \
　--password ADMIN_PASS admin

本书示例中将 admin 用户密码设置为 ADMIN_PASS，如图 5-8 所示。

```
[root@controller ~]# openstack user create --domain default \
> --password ADMIN_PASS admin
+-----------+----------------------------------+
| Field     | Value                            |
+-----------+----------------------------------+
| domain_id | 1224d01d12d5451191c2f0aeefa6ad6a |
| enabled   | True                             |
| id        | d8baad7133ec46caa2728a3610ee662a |
| name      | admin                            |
+-----------+----------------------------------+
```

图 5-8　创建 admin 用户

c. 创建 admin 角色：

\# openstack role create admin

如图 5-9 所示。

```
[root@controller ~]# openstack role create admin
+-----------+----------------------------------+
| Field     | Value                            |
+-----------+----------------------------------+
| domain_id | None                             |
| id        | 141cc8c0b5b8488fb27c001be03137d5 |
| name      | admin                            |
+-----------+----------------------------------+
```

图 5-9　创建 admin 角色

d. 添加 admin 角色到 admin 项目和用户上：

\# openstack role add--project admin--user admin admin

如图 5-10 所示。

```
[root@controller ~]# openstack role add --project admin --user admin admin
[root@controller ~]#
```

图 5 - 10　添加 admin 角色到 admin 项目和用户上

3）创建 service 项目，service 项目包含每个服务组件的唯一用户

openstack project create--domain default \

　　--description "Service Project" service

如图 5 - 11 所示。

```
[root@controller ~]# openstack project create --domain default \
> --description "Service Project" service
+-------------+----------------------------------+
| Field       | Value                            |
+-------------+----------------------------------+
| description | Service Project                  |
| domain_id   | 1224d01d12d5451191c2f0aeefa6ad6a |
| enabled     | True                             |
| id          | 3ef3b5ed56084a8b9edfdf1a679a822f |
| is_domain   | False                            |
| name        | service                          |
| parent_id   | 1224d01d12d5451191c2f0aeefa6ad6a |
+-------------+----------------------------------+
```

图 5 - 11　创建 service 项目

4）创建一个非管理员权限的项目、用户和角色，用于常规（非管理）任务操作

a. 创建 demo 项目：

openstack project create--domain default \

　　--description "Demo Project" demo

如图 5 - 12 所示。

```
[root@controller ~]# openstack project create --domain default \
> --description "Demo Project" demo
+-------------+----------------------------------+
| Field       | Value                            |
+-------------+----------------------------------+
| description | Demo Project                     |
| domain_id   | 1224d01d12d5451191c2f0aeefa6ad6a |
| enabled     | True                             |
| id          | 9ad59a65b5c44cc79146ef2c3b231c5e |
| is_domain   | False                            |
| name        | demo                             |
| parent_id   | 1224d01d12d5451191c2f0aeefa6ad6a |
+-------------+----------------------------------+
```

图 5 - 12　创建 demo 项目

b. 创建 demo 用户：

openstack user create--domain default \

　　--password DEMO_PASS demo

本书示例中将 demo 用户密码设置为 DEMO_PASS，如图 5-13 所示。

图 5-13　创建 demo 用户

c. 创建 user 角色：

openstack role create user

如图 5-14 所示。

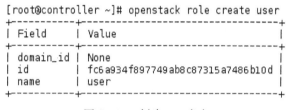

图 5-14　创建 user 角色

d. 添加 user 角色到 demo 项目和用户上：

openstack role add--project demo--user demo user

如图 5-15 所示。

```
[root@controller ~]# openstack role add --project demo --user demo user
[root@controller ~]#
```

图 5-15　添加 user 角色到 demo 项目和用户上

5.6　验证操作（控制节点）

1. 关闭临时认证令牌机制

出于安全性的原因，关闭临时认证令牌机制：

1）编辑 /etc/keystone/keystone-paste.ini 文件

gedit /etc/keystone/keystone-paste.ini

从 [pipeline：public_api]、[pipeline：admin_api]、[pipeline：api_v3] 部分，删除 admin_token_auth 字符：

[pipeline：public_api]

...

pipeline = cors sizelimit url_normalize request_id build_auth_context token_auth json_body ec2_extension public_service

[pipeline：admin_api]

...

pipeline = cors sizelimit url_normalize request_id build_auth_context token_auth json_body ec2_extension s3_extension admin_service

[pipeline：api_v3]

...

pipeline = cors sizelimit url_normalize request_id build_auth_context token_auth json_body ec2_extension_v3 s3_extension service_v3

2）重置 OS_TOKEN 和 OS_URL 环境变量

unset OS_TOKEN OS_URL

2. 使用 admin 用户请求认证令牌

这个命令使用 admin 用户的密码和 API 端口 35357，这样会允许对身份认证服务 API 的管理访问。

openstack--os-auth-url http：//controller：35357/v3 \

　　--os-project-domain-name default--os-user-domain-name default \

　　--os-project-name admin--os-username admin--os-password ADMIN_PASS \

　　token issue

本书示例中 admin 用户密码为 ADMIN_PASS，如图 5-16 所示。

```
[root@controller ~]# openstack --os-auth-url http://controller:35357/v3 \
> --os-project-domain-name default --os-user-domain-name default \
> --os-project-name admin --os-username admin --os-password ADMIN_PASS \
> token issue
+------------+-----------------------------------------------------------------+
| Field      | Value                                                           |
+------------+-----------------------------------------------------------------+
| expires    | 2017-03-27T03:10:07.133631Z                                     |
| id         | gAAAAABY2HR_7jkh1_yBLl6DnaFK2WB2P5j6o2BxicQe0hg-WCC-            |
|            | RLylinah0sE2wj9m5duX64WV0jRpj75mAspga9foElaCqsAYAS-             |
|            | mCa63xmy2y7snwZKu7I5IB7-i8zEIWnnb0vPqJZfHEZAB6Ubu4wz7MdVNQpagZiPkL2wBUZHItugXj1Y |
| project_id | cec93d07f74a4346a3b2221241abb52e                                |
| user_id    | d8baad7133ec46caa2728a3610ee662a                                |
+------------+-----------------------------------------------------------------+
```

图 5-16　使用 admin 用户请求认证令牌

3. 使用 demo 用户请求认证令牌

这个命令使用 demo 用户的密码和 API 端口 5000，这样只会允许对身份认证服务 API 的常规（非管理）访问。

openstack--os-auth-url http：//controller：5000/v3 \
　--os-project-domain-name default--os-user-domain-name default \
　--os-project-name demo--os-username demo--os-password DEMO_PASS \
　token issue

本书示例中 demo 用户密码为 DEMO_PASS，如图 5-17 所示。

```
[root@controller ~]# openstack --os-auth-url http://controller:5000/v3 \
> --os-project-domain-name default --os-user-domain-name default \
> --os-project-name demo --os-username demo --os-password DEMO_PASS \
> token issue
+------------+-------------------------------------------------------------------+
| Field      | Value                                                             |
+------------+-------------------------------------------------------------------+
| expires    | 2017-03-27T03:11:56.701648Z                                       |
| id         | gAAAAABY2HTsNAch7xCvrs-5NhNgVJmErSgaY3cEVmOrnG97MiAyMZT0V5__ZegSq01C8dNg4XNi7LWQ66 |
|            | b_gJz1iMeFaPogGdvacEpMdVvuTzvbN-xAI49QOE8CMOtX4x8KA22yZYREXoUEoaSQFK66IUxU-       |
|            | oh_JBOJBgvMyyqgNmXt6UsCyMY                                        |
| project_id | 9ad59a65b5c44cc79146ef2c3b231c5e                                  |
| user_id    | 455de29c0a3a4b86ac22b9f3f0b8ec67                                  |
+------------+-------------------------------------------------------------------+
```

图 5-17　使用 demo 用户请求认证令牌

5.7　创建 OpenStack 客户端环境脚本（控制节点）

1. 创建脚本

创建 admin 和 demo 项目和用户，创建客户端环境变量脚本。后续任务会引用这些脚本，为客户端操作加载合适的凭证。

1）创建 /root/admin-openrc 文件

gedit /root/admin-openrc

添加以下内容：

export OS_PROJECT_DOMAIN_NAME = default
export OS_USER_DOMAIN_NAME = default
export OS_PROJECT_NAME = admin
export OS_USERNAME = admin
export OS_PASSWORD = ADMIN_PASS
export OS_AUTH_URL = http：//controller：35357/v3
export OS_IDENTITY_API_VERSION = 3
export OS_IMAGE_API_VERSION = 2

2）创建 /root/demo-openrc 文件

gedit /root/demo-openrc

添加以下内容：

export OS_PROJECT_DOMAIN_NAME = default
export OS_USER_DOMAIN_NAME = default
export OS_PROJECT_NAME = demo
export OS_USERNAME = demo
export OS_PASSWORD = DEMO_PASS
export OS_AUTH_URL = http：//controller：5000/v3
export OS_IDENTITY_API_VERSION = 3
export OS_IMAGE_API_VERSION = 2

2. 使用脚本

使用特定用户运行客户端之前，可以加载相关客户端脚本。

1）加载 admin-openrc 文件来指定身份认证服务的环境变量位置和 admin 项目以及用户证书

./root/admin-openrc

2）请求认证令牌

openstack token issue

如图 5-18 所示。

```
[root@controller ~]# openstack token issue
+------------+-----------------------------------------------------------------+
| Field      | Value                                                           |
+------------+-----------------------------------------------------------------+
| expires    | 2017-03-27T03:12:54.685507Z                                     |
| id         | gAAAAABY2HUmaNZe0cFEDQ0OvcRwgF7S8uC5wWgRUigQ544SaFg5PC-0Nat2YY2mX8yNdCbc1FsL1Z- |
|            | C0bnpLg5hgTdyAVKjQJM2KHY4P4zUJVlaBGa49wxB4QS-                   |
|            | ffOdGsZhZ3-rycdIhSypwHcZxAomHTfhMiMx7nsPQ00YRdVvQOKAZXyklEQ     |
| project_id | cec93d07f74a4346a3b2221241abb52e                                |
| user_id    | d8baad7133ec46caa2728a3610ee662a                                |
+------------+-----------------------------------------------------------------+
```

图 5-18 请求认证令牌

第 6 章 镜像（Glance）服务

本章节示例将使用控制节点的本地文件系统中的一个目录（默认配置目录是/var/lib/glance/images/）作为后端配置镜像上传和存储的服务目录。

6.1 服务概述

OpenStack 镜像服务是基础设施即服务（IaaS）系统的核心服务。它接受磁盘镜像或服务器镜像 API 请求，来自终端用户或 OpenStack 计算组件的元数据定义。它也支持包括 OpenStack 对象存储在内的多种类型仓库上的磁盘镜像或服务器镜像存储。

Glance 镜像服务可实现发现、注册和获取虚拟机镜像和镜像元数据。通过 REST API 查询虚拟机镜像的 metadata（元数据）并获取一个现存的镜像。镜像数据文件可存储在多种存储系统中，例如简单的文件系统、对象存储系统等。

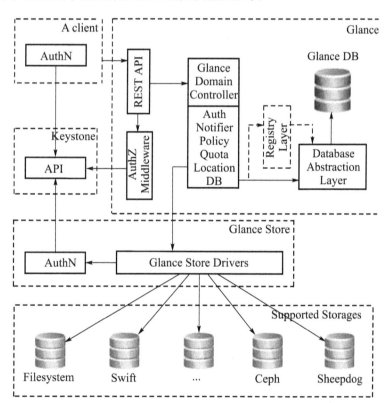

大量周期性进程运行于 OpenStack 镜像服务上以支持缓存同步复制（Replication）服务保证集群中的一致性和可用性。其他周期性进程还包括 auditors、updaters 和 reapers。

OpenStack 镜像服务包括以下组件：

1. glance-api

接收镜像 API 的调用，诸如镜像发现、恢复和存储。

2. glance-registry

存储、处理和恢复镜像的元数据，元数据项包括：诸如大小和类型。

3. Database（数据库）

存放镜像元数据，用户可以依据个人喜好选择数据库，多数的部署使用 MySQL 或 SQLite。

4. Storage repository for image files（镜像文件的存储仓库）

支持多种类型的仓库，它们有普通文件系统、对象存储、RADOS 块设备、HTTP 以及亚马逊 S3。注意其中一些仓库仅支持只读方式使用。

5. Metadata definition service（元数据定义服务）

通用的 API，为厂商、管理员、服务以及用户自定义元数据。这种元数据可用于不同的资源，例如镜像、工件、卷、配额以及集合。一个定义包括了新属性的键、描述、约束以及可以与之关联的资源的类型。

镜像文件格式：

在添加镜像到 Glance 时，需要指定虚拟镜像的磁盘格式（disk format）和容器格式（container format）。

虚拟镜像是一个虚拟机的磁盘映像，虚拟设备厂商将不同的格式布局的信息存在一个虚拟机磁盘映像文件中。

常见的磁盘格式有：raw、qcow2、AMI/AKI/ARI、VHD、VDI、VMDK、OVF。

容器格式是将虚拟机镜像添加元数据后重新打包形成的文件格式。

常见的容器格式有：Bare、ovf、aki、ari、ami、ova。

6.2 安装前准备（控制节点）

安装和配置镜像服务之前，必须创建一个数据库、服务凭证和 API 端点。

6.2.1 创建数据库

1）用数据库连接客户端以 root 用户连接到数据库服务器

mysql-u root-p'SQL_PASS'

2）创建 glance 数据库

CREATE DATABASE glance；

3）对 glance 数据库授予恰当的权限

本书示例中将 glance 数据库密码设置为 GLANCE_DBPASS，命令如下：

GRANT ALL PRIVILEGES ON glance.* TO 'glance'@'localhost' \
  IDENTIFIED BY 'GLANCE_DBPASS';

GRANT ALL PRIVILEGES ON glance.* TO 'glance'@'%' \
  IDENTIFIED BY 'GLANCE_DBPASS';

4）退出数据库客户端

exit

6.2.2 创建用户、服务实体和 API 端点

1. 获得 admin 凭证

获得 admin 凭证来获取只有管理员才能执行的命令的访问权限：

\# . /root/admin-openrc

2. 创建 glance 用户

1）创建 glance 用户

\# openstack user create --domain default --password GLANCE_PASS glance

本书示例中将 glance 用户密码设置为 GLANCE_PASS，如图 6-1 所示。

```
[root@controller ~]# openstack user create --domain default --password GLANCE_PASS glance
+-----------+----------------------------------+
| Field     | Value                            |
+-----------+----------------------------------+
| domain_id | 1224d01d12d5451191c2f0aeefa6ad6a |
| enabled   | True                             |
| id        | a89e297574b1495f98083d29ba480ce2 |
| name      | glance                           |
+-----------+----------------------------------+
```

图 6-1 创建 glance 用户

2）添加 admin 角色到 glance 用户和 service 项目上

\# openstack role add --project service --user glance admin

如图 6-2 所示。

```
[root@controller ~]# openstack role add --project service --user glance admin
[root@controller ~]#
```

图 6-2 添加 admin 角色到 glance 用户和 service 项目上

3. 创建服务实体和身份认证服务

创建 glance 服务实体:

openstack service create--name glance \
   --description "OpenStack Image" image

如图 6-3 所示。

```
[root@controller ~]# openstack service create --name glance \
> --description "OpenStack Image" image
+-------------+----------------------------------+
| Field       | Value                            |
+-------------+----------------------------------+
| description | OpenStack Image                  |
| enabled     | True                             |
| id          | b6440fca659944958dc88aa65645951f |
| name        | glance                           |
| type        | image                            |
+-------------+----------------------------------+
```

图 6-3　创建 glance 服务实体

4. 创建 API 端点

创建镜像服务的 API 端点:

openstack endpoint create--region RegionOne \
   image public http://controller:9292

创建 public 镜像服务的 API 端点, 如图 6-4 所示。

```
[root@controller ~]# openstack endpoint create --region RegionOne \
> image public http://controller:9292
+--------------+----------------------------------+
| Field        | Value                            |
+--------------+----------------------------------+
| enabled      | True                             |
| id           | 911626ee968f4a9cad3e6228777bfeb3 |
| interface    | public                           |
| region       | RegionOne                        |
| region_id    | RegionOne                        |
| service_id   | b6440fca659944958dc88aa65645951f |
| service_name | glance                           |
| service_type | image                            |
| url          | http://controller:9292           |
+--------------+----------------------------------+
```

图 6-4　创建镜像服务的 API 端点 (public)

openstack endpoint create--region RegionOne \
   image internal http://controller:9292

创建 internal 镜像服务的 API 端点, 如图 6-5 所示。

```
[root@controller ~]# openstack endpoint create --region RegionOne \
> image internal http://controller:9292
+--------------+----------------------------------+
| Field        | Value                            |
+--------------+----------------------------------+
| enabled      | True                             |
| id           | a5f29ddb1c1a4cf6ab019367d66f496e |
| interface    | internal                         |
| region       | RegionOne                        |
| region_id    | RegionOne                        |
| service_id   | b6440fca659944958dc88aa65645951f |
| service_name | glance                           |
| service_type | image                            |
| url          | http://controller:9292           |
+--------------+----------------------------------+
```

图 6-5 创建镜像服务的 API 端点（internal）

openstack endpoint create--region RegionOne \
　image admin http：//controller：9292

创建 admin 镜像服务的 API 端点，如图 6-6 所示。

```
[root@controller ~]# openstack endpoint create --region RegionOne \
> image admin http://controller:9292
+--------------+----------------------------------+
| Field        | Value                            |
+--------------+----------------------------------+
| enabled      | True                             |
| id           | e67de9472e2e453981faf7b7bf787946 |
| interface    | admin                            |
| region       | RegionOne                        |
| region_id    | RegionOne                        |
| service_id   | b6440fca659944958dc88aa65645951f |
| service_name | glance                           |
| service_type | image                            |
| url          | http://controller:9292           |
+--------------+----------------------------------+
```

图 6-6 创建镜像服务的 API 端点（admin）

6.3 安装和配置（控制节点）

6.3.1 安装软件包

安装 glance 软件包：
yum-y install openstack-glance

6.3.2 配置 Glance

1）编辑 /etc/glance/glance-api.conf 文件

由于默认配置文件在各发行版本中可能不同，因此，在进行修改的同时有可能需要添加部分选项。另外，在配置片段中的省略号（...）表示默认的配置选项，应该保留。

gedit /etc/glance/glance-api.conf

a. 在[database]部分，配置数据库访问：

[database]
...
connection = mysql + pymysql：//glance：GLANCE_DBPASS@ controller/glance

b. 在[keystone_authtoken]和[paste_deploy]部分，配置认证服务访问：

[keystone_authtoken]
...
auth_uri = http：//controller：5000
auth_url = http：//controller：35357
memcached_servers = controller：11211
auth_type = password
project_domain_name = default
user_domain_name = default
project_name = service
username = glance
password = GLANCE_PASS

[paste_deploy]
...
flavor = keystone

在[keystone_authtoken]中注释或者删除其他选项。

c. 在[glance_store]部分，配置本地文件系统存储和镜像文件存储位置：

[glance_store]
...
stores = file，http
default_store = file
filesystem_store_datadir = /var/lib/glance/images

d. 在[oslo_messaging_notifications]部分，配置信息通知：

[oslo_messaging_notifications]
...
driver = noop

2）编辑 /etc/glance/glance-registry.conf 文件

由于默认配置文件在各发行版本中可能不同，因此，在进行修改的同时可能需要添加部分选项。另外，在配置片段中的省略号（...）表示默认的配置选项，应该保留。

gedit /etc/glance/glance-registry.conf

a. 在［database］部分，配置数据库访问：

[database]
...
connection = mysql+pymysql：//glance：GLANCE_DBPASS@controller/glance

b. 在［keystone_authtoken］和［paste_deploy］部分，配置认证服务访问：

[keystone_authtoken]
...
auth_uri = http：//controller：5000
auth_url = http：//controller：35357
memcached_servers = controller：11211
auth_type = password
project_domain_name = default
user_domain_name = default
project_name = service
username = glance
password = GLANCE_PASS

[paste_deploy]
...
flavor = keystone

在［keystone_authtoken］中注释或者删除其他选项。

c. 在［oslo_messaging_notifications］部分，配置信息通知：

[oslo_messaging_notifications]
...
driver = noop

6.3.3 写入数据库

写入镜像服务数据库：

su -s /bin/sh -c "glance-manage db_sync" glance

如图 6-7 所示。

```
[root@controller ~]# su -s /bin/sh -c "glance-manage db_sync" glance
Option "verbose" from group "DEFAULT" is deprecated for removal.  Its value may be silently ignored
 in the future.
/usr/lib/python2.7/site-packages/oslo_db/sqlalchemy/enginefacade.py:1056: OsloDBDeprecationWarning:
 EngineFacade is deprecated; please use oslo_db.sqlalchemy.enginefacade
  expire_on_commit=expire_on_commit, _conf=conf)
/usr/lib/python2.7/site-packages/pymysql/cursors.py:166: Warning: (1831, u'Duplicate index 'ix_imag
e_properties_image_id_name' defined on the table 'glance.image_properties'. This is deprecated and
 will be disallowed in a future release.")
  result = self._query(query)
```

图 6-7 写入镜像服务数据库

忽略输出中任何不推荐使用的信息。

6.3.4 完成安装

启动镜像服务，并将其配置为随系统启动：

systemctl enable openstack-glance-api.service \
　　openstack-glance-registry.service

systemctl start openstack-glance-api.service \
　　openstack-glance-registry.service

6.4 验证操作（控制节点）

1）获得 admin 凭证来获取只有管理员才能执行的命令的访问权限

./root/admin-openrc

2）获得 CirrOS 源镜像

当主机可访问互联网时，可使用 wget 命令从在线源下载源镜像文件。

wget http：//download.cirros-cloud.net/0.3.4/cirros-0.3.4-x86_64-disk.img

在本书示例中将使用本地镜像文件 /openstack/data/cirros-0.3.4-x86_64-disk.img。

3）上传 CirrOS 镜像文件到镜像服务组件中

使用 QCOW2 磁盘格式时，将 bare 容器格式上传镜像到镜像服务并设置公共可见，这样所有的项目都可以访问它，如图 6-8 所示。

openstack image create "cirros" \
　　--file /openstack/data/cirros-0.3.4-x86_64-disk.img \

```
--disk-format qcow2 --container-format bare \
--public
```

```
[root@controller ~]# openstack image create "cirros" \
> --file /openstack/data/cirros-0.3.4-x86_64-disk.img \
> --disk-format qcow2 --container-format bare \
> --public
+------------------+------------------------------------------------------+
| Field            | Value                                                |
+------------------+------------------------------------------------------+
| checksum         | ee1eca47dc88f4879d8a229cc70a07c6                     |
| container_format | bare                                                 |
| created_at       | 2017-03-27T02:30:35Z                                 |
| disk_format      | qcow2                                                |
| file             | /v2/images/bbac27a8-868e-4b85-89d8-d14df3e9233f/file |
| id               | bbac27a8-868e-4b85-89d8-d14df3e9233f                 |
| min_disk         | 0                                                    |
| min_ram          | 0                                                    |
| name             | cirros                                               |
| owner            | cec93d07f74a4346a3b2221241abb52e                     |
| protected        | False                                                |
| schema           | /v2/schemas/image                                    |
| size             | 13287936                                             |
| status           | active                                               |
| tags             |                                                      |
| updated_at       | 2017-03-27T02:30:36Z                                 |
| virtual_size     | None                                                 |
| visibility       | public                                               |
+------------------+------------------------------------------------------+
```

图 6-8 上传 CirrOS 镜像文件

如果出现 HTTP 503 错误，那么请检查配置文件中的 glance 用户的密码是否正确，如图 6-9 所示。

```
[root@controller data]# openstack image create "cirros" \
> --file /openstack/data/cirros-0.3.4-x86_64-disk.img \
> --disk-format qcow2 --container-format bare \
> --public
503 Service Unavailable: The server is currently unavailable. Please try again at a later time. (HT
TP 503)
[root@controller data]# glance image-list
503 Service Unavailable: The server is currently unavailable. Please try again at a later time. (HT
TP 503)
```

图 6-9 HTTP 503 错误

4）确认镜像的上传并验证属性：
openstack image list

如图 6-10 所示。

```
[root@controller ~]# openstack image list
+--------------------------------------+--------+--------+
| ID                                   | Name   | Status |
+--------------------------------------+--------+--------+
| bbac27a8-868e-4b85-89d8-d14df3e9233f | cirros | active |
+--------------------------------------+--------+--------+
```

图 6-10 确认镜像的上传并验证属性

第 7 章　计算（Nova）服务

7.1　服务概述

　　OpenStack 计算服务是基础设施即服务（IaaS）系统的核心部分，模块主要由 Python 实现。OpenStack 计算服务用来托管和管理云计算系统，其功能包括运行虚拟机实例、管理网络以及通过用户和项目来控制对云的访问。

　　OpenStack 计算服务不包括任何的虚拟化软件。相反，它定义和运行在主机操作系统上虚拟化机制交互的驱动程序上，并通过基于 Web 的程序应用接口来提供功能的应用。

　　OpenStack 计算组件可以请求身份认证服务进行认证，请求镜像服务提供磁盘镜像，为仪表板服务提供用户与管理员接口。磁盘镜像访问限制在租户与用户上，而配额主要针对每个租户进行设定（例如，每个租户下可以创建多少实例）。

　　OpenStack **计算服务包括以下组件：**

　　1．nova-api 服务

　　Nova-api 是一个 HTTP 服务，用以接收和响应来自最终用户的计算 API 请求。它支持 OpenStackNova API、Amazon EC2 API 以及用于赋予用户做一些管理操作的特殊的管理 API。

　　2．nova-api-metadata 服务

　　接受来自虚拟机发送的元数据请求。Nova-api-metadata 服务一般在安装 Nova-network 服务的多主机模式下使用。

　　3．nova-compute 服务

　　Nova-compute 是 Nova 核心子组件，通过与 nova-client 进行交互，实现虚拟机的管理功能，负责在计算节点上对虚拟机实例进行一系列操作，包括迁移、安全组策略和快照管理等功能。

　　它是一个持续工作的守护进程，通过 Hypervior 的 API 来创建和销毁虚拟机实例。例如：

　　　　XenServer/XCP 的 Xen API

　　　　KVM 或 QEMU 的 libvirt

　　　　VMware 的 VMware API

守护进程同意了来自队列的动作请求之后，就转换为一系列的系统命令，如启动一个虚拟机实例，然后再到数据库中更新它的状态。

4．nova-scheduler 服务

完成 Nova 的核心调度，包括虚拟机硬件资源调度、节点调度等。同时，它还决定哪台计算服务器主机来运行虚拟机实例。

5．nova-conductor 模块

nova-conductor 模块是 OpenStack 中的一个 RPC 服务，媒介作用于 nova-compute 服务与数据库之间，避免了由 nova-compute 服务对云数据库的直接访问。

nova-conductor 模块可以水平扩展。但是不要将它部署在运行 nova-compute 服务的主机节点上。

6．nova-cert 模块

nova-cert 模块仅在 EC2 API 的请求中使用，服务器守护进程向 Nova Cert 服务提供 X509 证书，用来为 euca-bundle-image 生成证书。

7．nova-network worker 守护进程

与 nova-compute 服务类似，此进程从队列中接受网络任务，并且操作网络，同时执行任务，例如创建桥接的接口或者改变 IPtables 的规则。

8．nova-consoleauth 守护进程

授权 Nova 控制台代理（nova-novncproxy 和 nova-xvpvncproxy）所提供的用户令牌，实现对 Nova 控制台（VNC）的认证操作。该服务必须为控制台代理运行才可奏效。在集群配置中，用户可以运行二者中任一代理服务而非仅运行一个 nova-consoleauth 服务。

9．nova-novncproxy 守护进程

Nova 控制台组件，实现客户端与虚拟机实例的通信。它提供了一个 novnc 代理，用于访问正在运行的实例，并通过 VNC 协议来支持基于浏览器的 novnc 客户端。

10．nova-spicehtml5proxy 守护进程

Nova 控制台组件，它提供一个 html5 代理，用于访问正在运行的实例，并通过 SPICE 协议来支持基于浏览器的 html5 客户端。

11．nova-xvpvncproxy 守护进程

Nova 控制台组件，提供一个 VNC 代理，用于访问正在运行的实例，并通过 VNC 协议来支持 OpenStack 特定的 Java 客户端。

12．nova-cert 守护进程

用于管理提供 X509 认证证书，提供兼容性保障，保证所有的应用程序都能在云上

运行。

13. nova-client

nova-client 是一个命令行客户端应用，用于作为租户管理员或最终用户的人员提交命令。

14. 队列

一个在守护进程间传递消息的中央集线器。常见有 Rabbit MQ、Zero MQ 等 AMQP 消息队列。

15. SQL 数据库

存储构建时和运行时的状态，为云基础设施，包括有：

1）可用实例类型；
2）使用中的实例；
3）可用网络；
4）项目。

理论上，OpenStack 计算可以支持任何 SQL-Alchemy 所支持的后端数据库，通常使用 SQLite3 来做测试可开发工作，MySQL 和 PostgreSQL 作生产环境。

7.2 安装前准备（控制节点）

安装和配置计算服务之前，必须创建一个数据库、服务凭证和 API 端点。

7.2.1 创建数据库

1）用数据库连接客户端以 root 用户连接到数据库服务器

mysql -u root -p 'SQL_PASS'

2）创建 nova_api 和 nova 数据库

CREATE DATABASE nova_api;

CREATE DATABASE nova;

3）对 nova 数据库授予恰当的权限

本书示例中将 nova 数据库密码设置为 NOVA_DBPASS

GRANT ALL PRIVILEGES ON nova_api.* TO 'nova'@'localhost' \
　　IDENTIFIED BY 'NOVA_DBPASS';

GRANT ALL PRIVILEGES ON nova_api.* TO 'nova'@'%' \
　　IDENTIFIED BY 'NOVA_DBPASS';

GRANT ALL PRIVILEGES ON nova. * TO 'nova'@'localhost' \
    IDENTIFIED BY 'NOVA_DBPASS';
GRANT ALL PRIVILEGES ON nova. * TO 'nova'@'%' \
    IDENTIFIED BY 'NOVA_DBPASS';

4）退出数据库客户端

exit

7.2.2 创建用户、服务实体和 API 端点

1. 获得 admin 凭证

获得 admin 凭证来获取只有管理员才能执行的命令的访问权限：

#../root/admin-openrc

2. 创建 nova 用户

1）创建 nova 用户

openstack user create--domain default \
    --password NOVA_PASS nova

本书示例中将 nova 用户密码设置为 NOVA_PASS，如图 7-1 所示。

```
[root@controller ~]# openstack user create --domain default \
> --password NOVA_PASS nova
+-----------+----------------------------------+
| Field     | Value                            |
+-----------+----------------------------------+
| domain_id | 1224d01d12d5451191c2f0aeefa6ad6a |
| enabled   | True                             |
| id        | 060ac0850279406eb17511008848adef |
| name      | nova                             |
+-----------+----------------------------------+
```

图 7-1 创建 nova 用户

2）添加 admin 角色到 nova 用户和 service 项目上

openstack role add--project service--user nova admin

如图 7-2 所示。

```
[root@controller ~]# openstack role add --project service --user nova admin
[root@controller ~]#
```

图 7-2 添加 admin 角色到 nova 用户和 service 项目上

3. 创建服务实体和身份认证服务

创建 nova 服务实体：

openstack service create--name nova \
　　--description "OpenStack Compute" compute

如图 7-3 所示。

```
[root@controller ~]# openstack service create --name nova \
> --description "OpenStack Compute" compute
+-------------+----------------------------------+
| Field       | Value                            |
+-------------+----------------------------------+
| description | OpenStack Compute                |
| enabled     | True                             |
| id          | cb5b3c9faa0d45399a72d42b25bd2f8d |
| name        | nova                             |
| type        | compute                          |
+-------------+----------------------------------+
```

图 7-3　创建 nova 服务实体

4. 创建 API 端点

创建计算服务的 API 端点：

openstack endpoint create--region RegionOne \
　　compute public http：//controller：8774/v2.1/% \ （tenant_id \)s

创建 public、internal 和 admin 的 API 端点，分别如图 7-4、图 7-5 和图 7-6 所示。

```
[root@controller ~]# openstack endpoint create --region RegionOne \
> compute public http://controller:8774/v2.1/%\(tenant_id\)s
+--------------+----------------------------------+
| Field        | Value                            |
+--------------+----------------------------------+
| enabled      | True                             |
| id           | 7deb2da6f0e24bdaad78843525218341 |
| interface    | public                           |
| region       | RegionOne                        |
| region_id    | RegionOne                        |
| service_id   | cb5b3c9faa0d45399a72d42b25bd2f8d |
| service_name | nova                             |
| service_type | compute                          |
| url          | http://controller:8774/v2.1/%(tenant_id)s |
+--------------+----------------------------------+
```

图 7-4　创建计算服务的 API 端点（public）

openstack endpoint create--region RegionOne \
　　compute internal http：//controller：8774/v2.1/% \ （tenant_id \) s

```
[root@controller ~]# openstack endpoint create --region RegionOne \
> compute internal http://controller:8774/v2.1/%\(tenant_id\)s
+--------------+-------------------------------------------+
| Field        | Value                                     |
+--------------+-------------------------------------------+
| enabled      | True                                      |
| id           | ad906365d2eb4896bec40a67150f9dd3          |
| interface    | internal                                  |
| region       | RegionOne                                 |
| region_id    | RegionOne                                 |
| service_id   | cb5b3c9faa0d45399a72d42b25bd2f8d          |
| service_name | nova                                      |
| service_type | compute                                   |
| url          | http://controller:8774/v2.1/%(tenant_id)s |
+--------------+-------------------------------------------+
```

图 7-5 创建计算服务的 API 端点（internal）

openstack endpoint create--region RegionOne \
  compute admin http：//controller：8774/v2.1/% \ （tenant_id \）s

```
[root@controller ~]# openstack endpoint create --region RegionOne \
> compute admin http://controller:8774/v2.1/%\(tenant_id\)s
+--------------+-------------------------------------------+
| Field        | Value                                     |
+--------------+-------------------------------------------+
| enabled      | True                                      |
| id           | b97771891dfa46f7bd0213a1b33592ff          |
| interface    | admin                                     |
| region       | RegionOne                                 |
| region_id    | RegionOne                                 |
| service_id   | cb5b3c9faa0d45399a72d42b25bd2f8d          |
| service_name | nova                                      |
| service_type | compute                                   |
| url          | http://controller:8774/v2.1/%(tenant_id)s |
+--------------+-------------------------------------------+
```

图 7-6 创建计算服务的 API 端点（admin）

7.3 安装和配置（控制节点）

7.3.1 安装软件包

安装 nova 软件包：

yum -y install openstack-nova-api openstack-nova-conductor \
  openstack-nova-console openstack-nova-novncproxy \
  openstack-nova-scheduler

7.3.2 配置 Nova

编辑 /etc/nova/nova.conf 文件：

由于默认配置文件在各发行版本中可能不同，因此，在进行修改的同时可能需要添加部分选项。另外在配置片段中的省略号（...）表示默认的配置选项，应该保留。

gedit /etc/nova/nova.conf

a. 在 [DEFAULT] 部分，只启用计算和元数据 API：

[DEFAULT]
...
enabled_apis = osapi_compute, metadata

b. 在 [api_database] 和 [database] 部分，配置数据库的连接：

[api_database]
...
connection = mysql+pymysql://nova:NOVA_DBPASS@controller/nova_api

[database]
...
connection = mysql+pymysql://nova:NOVA_DBPASS@controller/nova

c. 在 [DEFAULT] 和 [oslo_messaging_rabbit] 部分，配置 RabbitMQ 消息队列访问：

[DEFAULT]
...
rpc_backend = rabbit

[oslo_messaging_rabbit]
...
rabbit_host = controller
rabbit_userid = openstack
rabbit_password = RABBIT_PASS

d. 在 [DEFAULT] 和 [keystone_authtoken] 部分，配置认证服务访问：

```
[DEFAULT]
...
auth_strategy = keystone

[keystone_authtoken]
...
auth_uri = http://controller:5000
auth_url = http://controller:35357
memcached_servers = controller:11211
auth_type = password
project_domain_name = default
user_domain_name = default
project_name = service
username = nova
password = NOVA_PASS
```

在［keystone_authtoken］中注释或者删除其他选项。

e. 在［DEFAULT］部分，配置 my_ip 的 IP 地址：

```
[DEFAULT]
...
my_ip = MANAGEMENT_INTERFACE_IP_ADDRESS
```

将 MANAGEMENT_INTERFACE_IP_ADDRESS 替换为控制节点的管理网络 IP 地址，本书示例中将其替换为 192.168.1.11。

f. 在［DEFAULT］部分，配置使用网络服务：

```
[DEFAULT]
...
use_neutron = True
firewall_driver = nova.virt.firewall.NoopFirewallDriver
```

默认情况下，计算服务使用内置的防火墙服务。由于网络服务包含了防火墙服务，因此，必须使用 nova.virt.firewall.NoopFirewallDriver 防火墙服务来禁用计算服务内置的防火墙服务。

g. 在［vnc］部分，配置 VNC 代理使用控制节点的管理接口 IP 地址：

[vnc]

...

vncserver_listen = $my_ip

vncserver_proxyclient_address = $my_ip

h. 在[glance]区域，配置镜像服务 API 的位置：

[glance]

...

api_servers = http://controller:9292

i. 在[oslo_concurrency]部分，配置锁路径：

[oslo_concurrency]

...

lock_path = /var/lib/nova/tmp

7.3.3 写入数据库

写入计算服务数据库：

su -s /bin/sh -c "nova-manage api_db sync" nova

su -s /bin/sh -c "nova-manage db sync" nova

如图 7-7 所示。

```
[root@controller ~]# su -s /bin/sh -c "nova-manage api_db sync" nova
[root@controller ~]# su -s /bin/sh -c "nova-manage db sync" nova
/usr/lib/python2.7/site-packages/pymysql/cursors.py:166: Warning: (1831, u'Duplicate index `block_d
evice_mapping_instance_uuid_virtual_name_device_name_idx`. This is deprecated and will be disallowe
d in a future release.')
  result = self._query(query)
/usr/lib/python2.7/site-packages/pymysql/cursors.py:166: Warning: (1831, u'Duplicate index `uniq_in
stances0uuid`. This is deprecated and will be disallowed in a future release.')
  result = self._query(query)
```

图 7-7 写入计算服务数据库

忽略输出中任何不推荐使用的信息。

7.3.4 完成安装

启动计算服务，并将其配置为随系统启动：

systemctl enable openstack-nova-api.service \

 openstack-nova-consoleauth.service openstack-nova-scheduler.service \

openstack-nova-conductor.service openstack-nova-novncproxy.service
systemctl start openstack-nova-api.service \
openstack-nova-consoleauth.service openstack-nova-scheduler.service \
openstack-nova-conductor.service openstack-nova-novncproxy.service

7.4 安装和配置（计算节点）

计算服务支持多种虚拟化方式，如 KVM、Xen、QEMU 和 VMware 等，可参照本节配置进行细微调整，以使用额外的计算节点横向扩展计算环境。每个额外的计算节点都需要一个唯一的 IP 地址。

7.4.1 安装软件包

安装 nova 软件包：
yum -y install openstack-nova-compute

7.4.2 配置 Nova

1）确认计算节点是否支持虚拟化硬件加速
egrep -c '(vmx|svm)' /proc/cpuinfo
此命令会返回一个数值，可通过此数值确认计算节点是否支持虚拟化硬件加速。
当命令返回数值大于或等于 1 时，表示计算节点支持虚拟化硬件加速：

```
[root@compute01 ~]# egrep -c '(vmx|svm)' /proc/cpuinfo
4
```

当命令返回数值等于 0 时，表示计算节点不支持虚拟化硬件加速。必须配置 libvirt 使用 QEMU 虚拟化方式来代替默认启用的 KVM 虚拟化方式：

```
[root@compute01 ~]# egrep -c '(vmx|svm)' /proc/cpuinfo
0
```

2）编辑 /etc/nova/nova.conf 文件
由于默认配置文件在各发行版本中可能不同，因此，在进行修改的同时可能需要添加部分选项。另外，在配置片段中的省略号（…）表示默认的配置选项，应该保留。
gedit /etc/nova/nova.conf
a. 在 [DEFAULT] 和 [oslo_messaging_rabbit] 部分，配置 RabbitMQ 消息队列访问：

```
[DEFAULT]
...
rpc_backend = rabbit

[oslo_messaging_rabbit]
...
rabbit_host = controller
rabbit_userid = openstack
rabbit_password = RABBIT_PASS
```

b. 在 [DEFAULT] 和 [keystone_authtoken] 部分，配置认证服务访问：

```
[DEFAULT]
...
auth_strategy = keystone

[keystone_authtoken]
...
auth_uri = http：//controller：5000
auth_url = http：//controller：35357
memcached_servers = controller：11211
auth_type = password
project_domain_name = default
user_domain_name = default
project_name = service
username = nova
password = NOVA_PASS
```

在 [keystone_authtoken] 中注释或者删除其他选项。

c. 在 [DEFAULT] 部分，配置 my_ip 的 IP 地址：

```
[DEFAULT]
...
my_ip = MANAGEMENT_INTERFACE_IP_ADDRESS
```

将 MANAGEMENT_INTERFACE_IP_ADDRESS 替换为计算节点的管理网络 IP 地址，本书示例中将其替换为 192.168.1.31。

d. 在[DEFAULT]部分，配置使用网络服务：

[DEFAULT]
...
use_neutron = True
firewall_driver = nova.virt.firewall.NoopFirewallDriver

默认情况下，计算服务使用内置的防火墙服务。由于网络服务包含了防火墙服务，因此，必须使用 nova.virt.firewall.NoopFirewallDriver 防火墙服务来禁用计算服务内置的防火墙服务。

e. 在[vnc]部分，启用并配置远程控制台访问：

[vnc]
...
enabled = True
vncserver_listen = 0.0.0.0
vncserver_proxyclient_address = $my_ip
novncproxy_base_url = http://controller:6080/vnc_auto.html

服务器组件监听所有的 IP 地址，而代理组件仅仅监听计算节点管理网络接口的 IP 地址。通过 URL 可以使用 Web 浏览器访问位于该计算节点上实例的远程控制台的位置。

如果运行浏览器的主机无法解析 controller 主机名，则可以将 controller 替换为控制节点管理网络的 IP 地址。

f. 在[glance]区域，配置镜像服务 API 的位置：

[glance]
...
api_servers = http://controller:9292

g. 在[oslo_concurrency]部分，配置锁路径：

[oslo_concurrency]
...
lock_path = /var/lib/nova/tmp

h. 在[DEFAULT]部分，配置 vif 插件：

[DEFAULT]
...
vif_plugging_is_fatal = false
vif_plugging_timeout = 10

ⅰ. 在 [libvirt] 部分，配置使用的虚拟化方式：

当计算节点支持虚拟化硬件加速时，使用 kvm 虚拟化方式。

[libvirt]
...
virt_type = kvm

当计算节点不支持虚拟化硬件加速时，使用 qemu 虚拟化方式。

[libvirt]
...
virt_type = qemu
cpu_mode = none

7.4.3　完成安装

启动计算服务，并将其配置为随系统启动：
systemctl enable libvirtd.service openstack-nova-compute.service
systemctl start libvirtd.service openstack-nova-compute.service

7.5　验证操作（控制节点）

1）获得 admin 凭证来获取只有管理员才能执行的命令的访问权限
./root/admin-openrc
2）列出服务组件，以验证是否成功启动并注册了每个进程
openstack compute service list
如图 7-8 所示。

```
[root@controller ~]# openstack compute service list
+----+------------------+------------+----------+---------+-------+----------------------------+
| Id | Binary           | Host       | Zone     | Status  | State | Updated At                 |
+----+------------------+------------+----------+---------+-------+----------------------------+
|  1 | nova-scheduler   | controller | internal | enabled | up    | 2017-03-27T03: 20: 25.000000 |
|  2 | nova-conductor   | controller | internal | enabled | up    | 2017-03-27T03: 20: 25.000000 |
|  3 | nova-consoleauth | controller | internal | enabled | up    | 2017-03-27T03: 20: 25.000000 |
|  7 | nova-compute     | compute01  | nova     | enabled | up    | 2017-03-27T03: 20: 27.000000 |
+----+------------------+------------+----------+---------+-------+----------------------------+
```

图 7-8 列出服务组件

在正确配置的情况下，该输出应该显示三个服务组件在控制节点上启用，一个服务组件在计算节点上启用。

第 8 章 网络（Neutron）服务

8.1 服务概述

OpenStack 网络服务允许创建、插入接口设备，这些设备由其他的 OpenStack 服务管理。插件式的实现可以容纳不同的网络设备和软件，为 OpenStack 架构与部署提供了灵活性。

OpenStack 网络主要和 OpenStack 计算交互，以提供网络连接到它的实例。

OpenStack 网络（neutron）管理 OpenStack 环境中所有虚拟网络基础设施（Virtual Networking Infrastructure，VNI）和物理网络基础设施（Physical Networking Infrastructure，PNI）的接入层。OpenStack 网络允许租户创建包括像 firewall、load balancer 和 virtual private networs（VPN）等在内的高级虚拟网络拓扑。

网络服务提供网络、子网和路由器等对象抽象，每个抽象都有模拟对应物理设备的功能：网络可包含子网，路由器在不同子网和网络间进行路由转发。

对于任意一个给定的网络都必须包含至少一个外部网络。不像其他的网络那样，外部网络不仅仅是一个定义的虚拟网络。相反，它代表了一种 OpenStack 安装之外的能从物理的、外部的网络访问的视图。外部网络上的 IP 地址可供外部网络上的任意的物理设备所访问外部网络之外，任何网络设置拥有一个或多个内部网络。因为这个网络仅仅代表外部网络的一部分，网络上 DHCP 是禁用的。

另外对应外部网络，任一网络服务设置有一个或以上的内部网络，这些软件定义的网络直接连接到虚拟机。只有当虚拟机在给定的内部网络或在通过类似路由器接口连接的子网，才能直接访问连接到该网络的虚拟机。

反之亦然，外网访问虚拟机需要路由器。每个路由器有一个连接到外部网络的网关和一个或多个连接内部网络的接口。类似物理路由器，子网可以访问连接到同一个路由器的其他子网上的设备，这些设备也能通过路由器的网关访问外网。

另外，外部网络的 IP 地址可以分配给内部网络端口。当有连接连接到子网时，该连接就被称为端口。通过关联外部网络 IP 地址和虚拟机端口，可使外网实体访问虚拟机。

网络服务也支持安全组。安全组允许管理员在组内定义防火墙规则。一个虚拟机可以属于一个或多个安全组，网络服务在这些安全组上应用这些规则来管理虚拟机端口的开与关、端口的范围和通信类型。

Neutron 网络服务组件是在 OpenStack Quantum 版本开始出现的，在此之前网络主要功能是通过 Nova-network 组件来实现，底层采用的大多是 Linux bridge，但是它无法快速组网和实现高级的网络功能，因此 OpenStack 把 Nova 中关于网络方面的功能进行了转移，成立了全新的 Neutron 组件，Neutron 组件取代了 Nova-network 的相关功能，但是 Nova 里还有些网络功能被保留，比如虚拟机的网卡方面的功能。

Neutron 其实是系统平台的位置，提供配置命令及参数检查，并把网络功能用一种逻辑组织起来，Neutron 自身并不提供任何网络功能，它只是一个框架，Neutron 的网络功能大部分是 Plugin（插件）提供的，Neutron 网络服务组件中的桥接技术（bridge agent）仅支持 VXLAN 重叠网络（Overlay Network）技术。

8.1.1 Neutron 网络服务组件支持的网络类型

1. Flat

所有虚拟机实例都连接在同一网络中，并且与物理宿主机运行在同一网络中，它不会对网络数据包标记 VLAN 标签或者隔离。

2. Local

虚拟机网络使用 Nova 计算服务组件的 nova-network，目前其代码已不再更新。由于它与 Neutron 网络服务组件功能重复，因此会被人为禁用。

3. VLAN

Neutron 网络服务组件使用 VLAN 标签（802.1Q tagged）技术允许用户创建多个公共网络和私有网络。虚拟机实例可以和网络中的所有设备进行通信，例如防火墙、Layer-2 交换机、负载均衡和服务器等。

4. GRE 和 VXLAN

GRE 和 VXLAN 是一种封装数据包的协议，创建重叠网络以激活和控制虚拟网络。Neutron 网络服务组件中的 neutron router 允许采用 GRE 和 VXLAN 的租户网络（私有网络）数据流出虚拟网络或租户网络，使数据自由流动。这是因为 neutron router 连接了租户网络（私有网络）和外部网络，使彼此互通。从外部网络访问租户网络（私有网络）中的虚拟机实例，需要使用浮动 IP（Floating IP）地址进行连接。

8.1.2 OpenStack 网络服务组件

1. neutron-server

接收和路由 API 请求到合适的 OpenStack 网络插件，以达到预想的目的。

2. OpenStack 网络插件和代理（OpenStack Networking plug-ins and agents）

可插拔端口，用于创建网络和子网以及提供 IP 地址，这些插件和代理依赖于供应商和技术而不同。OpenStack 网络基于插件和代理为 Cisco 虚拟和物理交换机、NEC OpenFlow 产品、Open vSwitch、Linux bridging 以及 VMware NSX 产品穿线搭桥。

常见的代理有 L3（3 层）、DHCP（动态主机 IP 地址）以及插件代理。

3. 消息队列（Messaging queue）

大多数的 OpenStack 网络安装都会用到消息队列，在 neutron-server 和各种各样的代理进程间路由信息，同时也为某些特定的插件扮演数据库的角色，以存储网络状态。

8.2 安装前准备（控制节点）

安装和配置网络服务之前，必须创建一个数据库、服务凭证和 API 端点。

8.2.1 创建数据库

1）用数据库连接客户端以 root 用户连接到数据库服务器
mysql-u root-p'SQL_PASS'
2）创建 neutron 数据库
CREATE DATABASE neutron;
3）对 neutron 数据库授予恰当的权限
本书示例中将 neutron 数据库密码设置为 NEUTRON_DBPASS。
GRANT ALL PRIVILEGES ON neutron.* TO 'neutron'@'localhost' \
 IDENTIFIED BY 'NEUTRON_DBPASS';
GRANT ALL PRIVILEGES ON neutron.* TO 'neutron'@'%' \
 IDENTIFIED BY 'NEUTRON_DBPASS';
4）退出数据库客户端
exit

8.2.2 创建用户、服务实体和 API 端点

1. 获得 admin 凭证

获得 admin 凭证来获取只有管理员才能执行的命令的访问权限：
./root/admin-openrc

2. 创建 neutron 用户

1）创建 neutron 用户
openstack user create--domain default \
 --password NEUTRON_PASS neutron
本书示例中将 neutron 用户密码设置为 NEUTRON_PASS，如图 8-1 所示。

```
[root@controller ~]# openstack user create --domain default \
> --password NEUTRON_PASS neutron
+-----------+----------------------------------+
| Field     | Value                            |
+-----------+----------------------------------+
| domain_id | 1224d01d12d5451191c2f0aeefa6ad6a |
| enabled   | True                             |
| id        | e60fce0c64b643fe9613cd1ca760a57f |
| name      | neutron                          |
+-----------+----------------------------------+
```

图 8-1　创建 neutron 用户

2) 添加 admin 角色到 neutron 用户和 service 项目上

openstack role add--project service--user neutron admin

如图 8-2 所示。

```
[root@controller ~]# openstack role add --project service --user neutron admin
[root@controller ~]#
```

图 8-2　添加 admin 角色到 neutron 用户及 service 项目上

3. 创建服务实体和身份认证服务

创建 neutron 服务实体：

openstack service create--name neutron \
　　--description "OpenStack Networking" network

如图 8-3 所示。

```
[root@controller ~]# openstack service create --name neutron \
> --description "OpenStack Networking" network
+-------------+----------------------------------+
| Field       | Value                            |
+-------------+----------------------------------+
| description | OpenStack Networking             |
| enabled     | True                             |
| id          | f6e98a9552bb456dba1ceca74fe12cf3 |
| name        | neutron                          |
| type        | network                          |
+-------------+----------------------------------+
```

图 8-3　创建 neutron 服务实体

4. 创建 API 端点

创建网络服务的 API 端点：

openstack endpoint create--region RegionOne \

network public http：//controller：9696

创建 public API 端点,如图 8-4 所示。

```
[root@controller ~]# openstack endpoint create --region RegionOne \
> network public http://controller:9696
+--------------+----------------------------------+
| Field        | Value                            |
+--------------+----------------------------------+
| enabled      | True                             |
| id           | c36ba32765224b779d65d2cae565226b |
| interface    | public                           |
| region       | RegionOne                        |
| region_id    | RegionOne                        |
| service_id   | f6e98a9552bb456dba1ceca74fe12cf3 |
| service_name | neutron                          |
| service_type | network                          |
| url          | http://controller:9696           |
+--------------+----------------------------------+
```

图 8-4 创建网络服务的 API 端点 (public)

openstack endpoint create--region RegionOne \

network internal http：//controller：9696

创建 internal API 端点,如图 8-5 所示。

```
[root@controller ~]# openstack endpoint create --region RegionOne \
> network internal http://controller:9696
+--------------+----------------------------------+
| Field        | Value                            |
+--------------+----------------------------------+
| enabled      | True                             |
| id           | 2fad6374aee44b6cbaf025ffd47079b9 |
| interface    | internal                         |
| region       | RegionOne                        |
| region_id    | RegionOne                        |
| service_id   | f6e98a9552bb456dba1ceca74fe12cf3 |
| service_name | neutron                          |
| service_type | network                          |
| url          | http://controller:9696           |
+--------------+----------------------------------+
```

图 8-5 创建网络服务的 API 端点 (internal)

openstack endpoint create--region RegionOne \

network admin http：//controller：9696

创建 admin API 端点,如图 8-6 所示。

```
[root@controller ~]# openstack endpoint create --region RegionOne \
> network admin http://controller:9696
+--------------+----------------------------------+
| Field        | Value                            |
+--------------+----------------------------------+
| enabled      | True                             |
| id           | 040f7ed1ba524ff08956f646ceb08bac |
| interface    | admin                            |
| region       | RegionOne                        |
| region_id    | RegionOne                        |
| service_id   | f6e98a9552bb456dba1ceca74fe12cf3 |
| service_name | neutron                          |
| service_type | network                          |
| url          | http://controller:9696           |
+--------------+----------------------------------+
```

图 8-6 创建网络服务的 API 端点 (admin)

8.3 安装和配置（控制节点）

OpenStack 平台中的网络分为两类，分别是公共网络（Provider networks）和私有网络（Self-service networks）。

私有网络（Self-service networks）完全兼容公共网络（Provider networks）功能并进行了扩展。增加支持 Layer-3（路由）服务和 VXLAN 重叠网络技术。

公共网络（Provider networks）采用尽可能简单的架构进行部署，只支持实例连接到公共网络（外部网络）。没有私有网络（个人网络）、路由器以及浮动 IP 地址。只有 admin 或者其他特权用户才可以管理公共网络。

私有网络（Self-service networks）在公共网络（Provider networks）的基础上多了 Layer-3 服务，支持实例连接到私有网络。demo 或者其他没有特权的用户可以管理自己的私有网络，包含连接公共网络和私有网络的路由器。另外，浮动 IP 地址可以让实例使用私有网络连接到外部网络，例如互联网。

典型的私有网络一般使用重叠网络。例如 VXLAN 包含了额外的数据头，这些数据头增加了开销，减少了有效内容和用户数据的可用空间。在不了解虚拟网络架构的情况下，实例尝试用以太网最大传输单元（MTU）1500 字节发送数据包。网络服务通过 DHCP 的方式自动给实例提供正确的 MTU 值。但是，一些云镜像并没有使用 DHCP 或者忽视了 DHCP MTU 选项，要求使用元数据或者脚本来进行配置。

公共网络（Provider networks）和私有网络（Self-service networks）的配置过程会存在一些差异，请留意步骤中的相关说明。

本书示例中将以私有网络（Self-service networks）类型作为基础网络，因此请按私有网络（Self-service networks）的配置步骤进行操作。

8.3.1 安装软件包

安装 neutron 软件包：
yum -y install openstack-neutron openstack-neutron-ml2 \
　　openstack-neutron-linuxbridge ebtables

8.3.2 配置 Neutron

Networking 服务器组件的配置包括数据库、认证机制、消息队列、拓扑变化通知和插件。

本书示例中使用私有网络（Self-service networks）的配置。

编辑 /etc/neutron/neutron.conf 文件：

由于默认配置文件在各发行版本中可能不同，因此，在进行修改的同时可能需要添加部分选项。另外，在配置片段中的省略号（...）表示默认的配置选项，应该保留。

gedit /etc/neutron/neutron.conf

a. 在 [database] 部分，配置数据库访问：

[database]
...
connection = mysql+pymysql://neutron:NEUTRON_DBPASS@controller/neutron

b. 在 [DEFAULT] 部分，启用 ML2 插件并禁用其他插件：

(a) 公共网络（Provider networks）启用 ML2 插件并禁用其他插件：

[DEFAULT]
...
core_plugin = ml2
service_plugins =

(b) 私有网络（Self-service networks）启用 ML2 插件、路由服务和重叠 IP 地址：

[DEFAULT]
...
core_plugin = ml2
service_plugins = router
allow_overlapping_ips = True

c. 在 [DEFAULT] 和 [oslo_messaging_rabbit] 部分，配置 RabbitMQ 消息队列访问：

```
[DEFAULT]
...
rpc_backend = rabbit

[oslo_messaging_rabbit]
...
rabbit_host = controller
rabbit_userid = openstack
rabbit_password = RABBIT_PASS
```

d. 在 [DEFAULT] 和 [keystone_authtoken] 部分，配置认证服务访问：

```
[DEFAULT]
...
auth_strategy = keystone

[keystone_authtoken]
...
auth_uri = http://controller:5000
auth_url = http://controller:35357
memcached_servers = controller:11211
auth_type = password
project_domain_name = default
user_domain_name = default
project_name = service
username = neutron
password = NEUTRON_PASS
```

在 [keystone_authtoken] 中注释或者删除其他选项。

e. 在 [DEFAULT] 和 [nova] 部分，配置网络服务来通知计算节点的网络拓扑变化：

```
[DEFAULT]
...
notify_nova_on_port_status_changes = True
notify_nova_on_port_data_changes = True

[nova]
```

```
...
region_name = RegionOne
auth_url = http://controller:35357
auth_type = password
project_domain_name = default
project_name = service
user_domain_name = default
username = nova
password = NOVA_PASS
```

f. 在[oslo_concurrency]部分，配置锁路径：

```
[oslo_concurrency]
...
lock_path = /var/lib/neutron/tmp
```

8.3.3 配置 ML2 插件

Modular Layer 2 (ML2) 插件使用 Linuxbridge 机制来为实例创建 Layer-2 虚拟网络基础设施。

本书示例中使用私有网络（Self-service networks）的配置。

编辑 /etc/neutron/plugins/ml2/ml2_conf.ini 文件：

由于默认配置文件在各发行版本中可能不同，因此，在进行修改的同时可能需要添加部分选项。另外，在配置片段中的省略号（...）表示默认的配置选项，应该保留。

gedit /etc/neutron/plugins/ml2/ml2_conf.ini

a. 在[ml2]部分，配置要启用的网络类型：

(a) 公共网络（Provider networks）启用 flat 和 VLAN 网络：

```
[ml2]
...
type_drivers = flat, vlan
```

(b) 私有网络（Self-service networks）启用 flat、VLAN 和 VXLAN 网络：

```
[ml2]
...
type_drivers = flat, vlan, vxlan
```

ML2 插件配置完成后，请不要删除 type_drivers 项的值。因为删除可能会导致数据库中 type_drivers 项的值不一致。

b. 在［ml2］部分，配置租户网络类型：

（a）公共网络（Provider networks）禁用租户网络：

［ml2］
...
tenant_network_types =

（b）私有网络（Self-service networks）启用 VXLAN 重叠网络：

［ml2］
...
tenant_network_types = vxlan

Linuxbridge 代理只支持 VXLAN 重叠网络。

c. 在［ml2］部分，配置驱动机制：

（a）公共网络（Provider networks）启用 Linuxbridge 机制：

［ml2］
...
mechanism_drivers = linuxbridge

（b）私有网络（Self-service networks）启用 Linuxbridge 和 Layer-2 机制：

［ml2］
...
mechanism_drivers = linuxbridge，l2population

d. 在［ml2］部分，启用端口安全扩展驱动：

［ml2］
...
extension_drivers = port_security

e. 在［ml2_type_flat］部分，配置 flat 网络为公共虚拟网络：

［ml2_type_flat］
...
flat_networks = provider

f. 在［ml2_type_vxlan］部分，配置 VXLAN 网络的识别范围：
（a）公共网络（Provider networks）不支持 VXLAN 重叠网络，无须进行配置：

［ml2_type_vxlan］
...
#vni_ranges =

（b）私有网络（Self-service networks）配置 VXLAN 重叠网络的识别范围：

［ml2_type_vxlan］
...
vni_ranges = 1：1000

g. 在［securitygroup］部分，启用 ipset 增加安全组规则的高效性：

［securitygroup］
...
enable_ipset = True

8.3.4 配置 Linuxbridge 代理

Linuxbridge 代理为实例建立 Layer-2 虚拟网络并且处理安全组规则。
本书示例中使用私有网络（Self-service networks）的配置。
编辑 /etc/neutron/plugins/ml2/linuxbridge_agent.ini 文件：
由于默认配置文件在各发行版本中可能不同，因此，在进行修改的同时可能需要添加部分选项。另外，在配置片段中的省略号（...）表示默认的配置选项，应该保留。
gedit /etc/neutron/plugins/ml2/linuxbridge_agent.ini
a. 在［linux_bridge］部分，将公共虚拟网络和公共物理网络接口对应起来：

［linux_bridge］
...
physical_interface_mappings = provider：PROVIDER_INTERFACE_NAME

将 PROVIDER_INTERFACE_NAME 替换为连接到公共网络的物理接口的设备名称。
由于本书示例中使用主机中的第 2 块网卡作为连接到公共网络的物理接口，因此，需将 PROVIDER_INTERFACE_NAME 替换为控制节点的第 2 块网卡接口的设备名称 eno33557248。
b. 在［vxlan］部分，配置 VXLAN 重叠网络：
（a）公共网络（Provider networks）禁用 VXLAN 重叠网络：

```
[vxlan]
...
enable_vxlan = false
```

(b) 私有网络（Self-service networks）启用 VXLAN 重叠网络，配置重叠网络的物理网络接口的 IP 地址，启用 Layer-2 通道：

```
[vxlan]
...
enable_vxlan = True
local_ip = OVERLAY_INTERFACE_IP_ADDRESS
l2_population = True
```

将 OVERLAY_INTERFACE_IP_ADDRESS 替换为处理重叠网络的底层物理网络接口的 IP 地址。

由于本书示例中使用管理网络接口与其他节点建立流量隧道，因此，需将 OVERLAY_INTERFACE_IP_ADDRESS 替换为控制节点的管理网络 IP 地址 192.168.1.11。

c. 在 [securitygroup] 部分，启用安全组并配置 Linuxbridge iptables firewall driver：

```
[securitygroup]
...
firewall_driver = neutron.agent.linux.iptables_firewall.IptablesFirewallDriver
enable_security_group = True
```

8.3.5 配置 DHCP 代理

使用 DHCP 代理为虚拟网络提供 DHCP 服务，在公共网络上的实例可以通过网络来访问元数据。

编辑 /etc/neutron/dhcp_agent.ini 文件：

由于默认配置文件在各发行版本中可能不同，因此，在进行修改的同时可能需要添加部分选项。另外，在配置片段中的省略号（...）表示默认的配置选项，应该保留。

gedit /etc/neutron/dhcp_agent.ini

在 [DEFAULT] 部分，配置 Linuxbridge 驱动接口、DHCP 驱动并启用隔离元数据。

[DEFAULT]
...
interface_driver = neutron.agent.linux.interface.BridgeInterfaceDriver
dhcp_driver = neutron.agent.linux.dhcp.Dnsmasq
enable_isolated_metadata = True

8.3.6 配置 Layer-3 代理

只有私有网络（Self-service networks）才需配置使用 Layer-3 代理。

Layer-3 代理为私有网络（Self-service networks）提供路由和 NAT 服务。

本书示例中使用私有网络（Self-service networks）的配置。

编辑 /etc/neutron/l3_agent.ini 文件：

由于默认配置文件在各发行版本中可能不同，因此，在进行修改的同时可能需要添加部分选项。另外，在配置片段中的省略号（...）表示默认的配置选项，应该保留。

gedit /etc/neutron/l3_agent.ini

在［DEFAULT］部分，配置 Linuxbridge 接口驱动和外部网络网桥。

[DEFAULT]
...
interface_driver = neutron.agent.linux.interface.BridgeInterfaceDriver
external_network_bridge =

将 external_network_bridge 选项特意设置成缺省值，这样就可以在一个代理上允许多种外部网络。

8.3.7 配置元数据代理

Metadata agent 负责提供配置信息，例如访问实例的凭证。

编辑 /etc/neutron/metadata_agent.ini 文件：

由于默认配置文件在各发行版本中可能不同，因此，在进行修改的同时可能需要添加部分选项。另外，在配置片段中的省略号（...）表示默认的配置选项，应该保留。

gedit /etc/neutron/metadata_agent.ini

在［DEFAULT］部分，配置元数据主机以及共享密码：

```
[DEFAULT]
...
nova_metadata_ip = controller
metadata_proxy_shared_secret = METADATA_SECRET
```

本书示例中将元数据代理的共享密码设置为 METADATA_SECRET。

8.3.8　为计算服务配置网络服务

编辑 /etc/nova/nova.conf 文件：

由于默认配置文件在各发行版本中可能不同，因此，在进行修改的同时可能需要添加部分选项。另外，在配置片段中的省略号（...）表示默认的配置选项，应该保留。

gedit /etc/nova/nova.conf

在［neutron］部分，配置访问参数，启用元数据代理并设置密码：

```
[neutron]
...
service_metadata_proxy = True
metadata_proxy_shared_secret = METADATA_SECRET

url = http://controller:9696
region_name = RegionOne
auth_url = http://controller:35357
auth_type = password
project_domain_name = default
project_name = service
user_domain_name = default
username = neutron
password = NEUTRON_PASS
```

本书示例中将元数据代理的共享密码设置为 METADATA_SECRET。

8.3.9　写入数据库

1）创建网络服务初始化脚本超链接

网络服务初始化脚本需要一个超链接 /etc/neutron/plugin.ini 指向 ML2 插件配置文件 /etc/neutron/plugins/ml2/ml2_conf.ini。如果超链接不存在，则使用下面的命令创建它：

ln -s /etc/neutron/plugins/ml2/ml2_conf.ini /etc/neutron/plugin.ini

2）写入网络服务数据库

su-s /bin/sh -c "neutron-db-manage --config-file /etc/neutron/neutron.conf \
    --config-file /etc/neutron/plugins/ml2/ml2_conf.ini upgrade head" neutron

如图 8-7 所示。

```
[root@controller ~]# su -s /bin/sh -c "neutron-db-manage --config-file /etc/neutron/neutron.conf \
> --config-file /etc/neutron/plugins/ml2/ml2_conf.ini upgrade head" neutron
No handlers could be found for logger "oslo_config.cfg"
INFO  [alembic.runtime.migration] Context impl MySQLImpl.
INFO  [alembic.runtime.migration] Will assume non-transactional DDL.
  Running upgrade for neutron ...
INFO  [alembic.runtime.migration] Context impl MySQLImpl.
INFO  [alembic.runtime.migration] Will assume non-transactional DDL.
INFO  [alembic.runtime.migration] Running upgrade  -> kilo, kilo_initial
INFO  [alembic.runtime.migration] Running upgrade kilo -> 354db87e3225, nsxv_vdr_metadata.py
INFO  [alembic.runtime.migration] Running upgrade 354db87e3225 -> 599c6a226151, neutrodb_ipam
INFO  [alembic.runtime.migration] Running upgrade 599c6a226151 -> 52c5312f6baf, Initial operations in support of address scopes
```

图 8-7　写入网络服务数据库

网络服务数据库的写入将发生在网络服务初始化之后，因为网络服务初始化脚本需要完成服务器和插件的配置文件。

8.3.10　完成安装

1）重启计算 API 服务

systemctl restart openstack-nova-api.service

2）启动网络服务，并将其配置为随系统启动

systemctl enable neutron-server.service \
    neutron-linuxbridge-agent.service neutron-dhcp-agent.service \
    neutron-metadata-agent.service

systemctl start neutron-server.service \
    neutron-linuxbridge-agent.service neutron-dhcp-agent.service \
    neutron-metadata-agent.service

3）为私有网络（Self-service networks）启动 Layer-3 服务，并将其配置为随系统启动

systemctl enable neutron-l3-agent.service

systemctl start neutron-l3-agent.service

8.4　安装和配置（计算节点）

在计算节点上，网络通用组件服务用于处理实例的连接和安全组。

8.4.1 安装软件包

安装 Neutron 通用组件软件包:
yum -y install openstack-neutron-linuxbridge ebtables ipset

8.4.2 配置 Neutron

编辑 /etc/neutron/neutron.conf 文件:

由于默认配置文件在各发行版本中可能不同,因此,在进行修改的同时可能需要添加部分选项。另外,在配置片段中的省略号(...)表示默认的配置选项,应该保留。
gedit /etc/neutron/neutron.conf

a. 在[database]部分,注释掉所有 connection 项,因为计算节点不直接访问数据库。

b. 在[DEFAULT]和[oslo_messaging_rabbit]部分,配置 RabbitMQ 消息队列访问:

[DEFAULT]
...
rpc_backend = rabbit

[oslo_messaging_rabbit]
...
rabbit_host = controller
rabbit_userid = openstack
rabbit_password = RABBIT_PASS

c. 在[DEFAULT]和[keystone_authtoken]部分,配置认证服务访问:

[DEFAULT]
...
auth_strategy = keystone

[keystone_authtoken]
...
auth_uri = http://controller:5000
auth_url = http://controller:35357
memcached_servers = controller:11211
auth_type = password
project_domain_name = default
user_domain_name = default

```
project_name = service
username = neutron
password = NEUTRON_PASS
```

在［keystone_authtoken］中注释或者删除其他选项。

d. 在［oslo_concurrency］部分，配置锁路径：

```
[oslo_concurrency]
...
lock_path = /var/lib/neutron/tmp
```

8.4.3 配置 Linuxbridge 代理

Linuxbridge 代理为实例建立 Layer-2 虚拟网络并且处理安全组规则。

编辑 /etc/neutron/plugins/ml2/linuxbridge_agent.ini 文件：

由于默认配置文件在各发行版本中可能不同，因此，在进行修改的同时可能需要添加部分选项。另外，在配置片段中的省略号（...）表示默认的配置选项，应该保留。

`# gedit /etc/neutron/plugins/ml2/linuxbridge_agent.ini`

a. 在［linux_bridge］部分，将公共虚拟网络和公共物理网络接口对应起来：

```
[linux_bridge]
...
physical_interface_mappings = provider：PROVIDER_INTERFACE_NAME
```

将 PROVIDER_INTERFACE_NAME 替换为连接到公共网络的物理接口的设备名称。

由于本书示例中使用主机中的第 2 块网卡作为连接到公共网络的物理接口，因此，需要将 PROVIDER_INTERFACE_NAME 替换为计算节点的第 2 块网卡接口的设备名称 eno33557248。

b. 在［vxlan］部分，配置 VXLAN 重叠网络：

（a）公共网络（Provider networks）禁用 VXLAN 重叠网络：

```
[vxlan]
...
enable_vxlan = false
```

（b）私有网络（Self-service networks）启用 VXLAN 重叠网络，配置重叠网络的物理网络接口的 IP 地址，启用 Layer-2 通道：

```
[vxlan]
...
enable_vxlan = True
local_ip = OVERLAY_INTERFACE_IP_ADDRESS
l2_population = True
```

将 OVERLAY_INTERFACE_IP_ADDRESS 替换为处理重叠网络的底层物理网络接口的 IP 地址。

由于本书示例中使用管理网络接口与其他节点建立流量隧道，因此，需将 OVERLAY_INTERFACE_IP_ADDRESS 替换为计算节点的管理网络 IP 地址 192.168.1.31。

c. 在 [securitygroup] 部分，启用安全组并配置 Linuxbridge iptables firewall driver：

```
[securitygroup]
...
firewall_driver = neutron.agent.linux.iptables_firewall.IptablesFirewallDriver
enable_security_group = True
```

8.4.4 为计算服务配置网络服务

编辑 /etc/nova/nova.conf 文件：

由于默认配置文件在各发行版本中可能不同，因此，在进行修改的同时可能需要添加部分选项。另外，在配置片段中的省略号（...）表示默认的配置选项，应该保留。

gedit /etc/nova/nova.conf

在 [neutron] 部分，配置访问参数：

```
[neutron]
...
url = http://controller:9696
region_name = RegionOne
auth_url = http://controller:35357
auth_type = password
project_domain_name = default
project_name = service
user_domain_name = default
username = neutron
password = NEUTRON_PASS
```

8.4.5 完成安装

1）重启计算服务

\# systemctl restart openstack-nova-compute. service

2）启动 Linuxbridge 代理服务，并将其配置为随系统启动

\# systemctl enable neutron-linuxbridge-agent. service

\# systemctl start neutron-linuxbridge-agent. service

8.5 验证操作（控制节点）

1）获得 admin 凭证来获取只有管理员才能执行的命令的访问权限

\# ./root/admin-openrc

2）列出网络服务扩展模块，以验证 neutron 服务进程是否正常启动

\# neutron ext-list

a. 公共网络（Provider networks），如图 8-8 所示。

```
[root@controller ~]# neutron ext-list
+------------------------------+----------------------------------------------+
| alias                        | name                                         |
+------------------------------+----------------------------------------------+
| default-subnetpools          | Default Subnetpools                          |
| availability_zone            | Availability Zone                            |
| network_availability_zone    | Network Availability Zone                    |
| auto-allocated-topology      | Auto Allocated Topology Services             |
| binding                      | Port Binding                                 |
| agent                        | agent                                        |
| subnet_allocation            | Subnet Allocation                            |
| dhcp_agent_scheduler         | DHCP Agent Scheduler                         |
| tag                          | Tag support                                  |
| external-net                 | Neutron external network                     |
| net-mtu                      | Network MTU                                  |
| network-ip-availability      | Network IP Availability                      |
| quotas                       | Quota management support                     |
| provider                     | Provider Network                             |
| multi-provider               | Multi Provider Network                       |
| address-scope                | Address scope                                |
| timestamp_core               | Time Stamp Fields addition for core resources|
| extra_dhcp_opt               | Neutron Extra DHCP opts                      |
| security-group               | security-group                               |
| rbac-policies                | RBAC Policies                                |
| standard-attr-description    | standard-attr-description                    |
| port-security                | Port Security                                |
| allowed-address-pairs        | Allowed Address Pairs                        |
+------------------------------+----------------------------------------------+
```

图 8-8 列出公共网络服务扩展模块

b. 私有网络（Self-service networks），如图 8-9 所示。

```
[root@controller ~]# neutron ext-list
+-----------------------------+------------------------------------------------+
| alias                       | name                                           |
+-----------------------------+------------------------------------------------+
| default-subnetpools         | Default Subnetpools                            |
| network-ip-availability     | Network IP Availability                        |
| network_availability_zone   | Network Availability Zone                      |
| auto-allocated-topology     | Auto Allocated Topology Services               |
| ext-gw-mode                 | Neutron L3 Configurable external gateway mode  |
| binding                     | Port Binding                                   |
| agent                       | agent                                          |
| subnet_allocation           | Subnet Allocation                              |
| l3_agent_scheduler          | L3 Agent Scheduler                             |
| tag                         | Tag support                                    |
| external-net                | Neutron external network                       |
| net-mtu                     | Network MTU                                    |
| availability_zone           | Availability Zone                              |
| quotas                      | Quota management support                       |
| l3-ha                       | HA Router extension                            |
| provider                    | Provider Network                               |
| multi-provider              | Multi Provider Network                         |
| address-scope               | Address scope                                  |
| extraroute                  | Neutron Extra Route                            |
| timestamp_core              | Time Stamp Fields addition for core resources  |
| router                      | Neutron L3 Router                              |
| extra_dhcp_opt              | Neutron Extra DHCP opts                        |
| security-group              | security-group                                 |
| dhcp_agent_scheduler        | DHCP Agent Scheduler                           |
| router_availability_zone    | Router Availability Zone                       |
| rbac-policies               | RBAC Policies                                  |
| standard-attr-description   | standard-attr-description                      |
| port-security               | Port Security                                  |
| allowed-address-pairs       | Allowed Address Pairs                          |
| dvr                         | Distributed Virtual Router                     |
+-----------------------------+------------------------------------------------+
```

图8-9 列出私有网络服务扩展模块

3) 列出网络服务代理,以验证 neutron 代理是否启动成功

neutron agent-list

a. 公共网络（Provider networks）,如图8-10所示。

```
[root@controller ~]# neutron agent-list
+------------------+--------------------+------------+-------------------+-------+----------------+------------------------+
| id               | agent_type         | host       | availability_zone | alive | admin_state_up | binary                 |
+------------------+--------------------+------------+-------------------+-------+----------------+------------------------+
| 5ff4b987         | Metadata           | controller |                   | :-)   | True           | neutron-               |
| -f35f-40ed-9a    | agent              |            |                   |       |                | metadata-              |
| 58-d6ef7550ec    |                    |            |                   |       |                | agent                  |
| b2               |                    |            |                   |       |                |                        |
| 7e2cef53         | Linux bridge       | controller |                   | :-)   | True           | neutron-               |
| -ee4c-4c98-86    | agent              |            |                   |       |                | linuxbridge-           |
| d4-ab9fbf3532    |                    |            |                   |       |                | agent                  |
| 10               |                    |            |                   |       |                |                        |
| b66555bb-        | DHCP agent         | controller | nova              | :-)   | True           | neutron-dhcp-          |
| a2eb-40ef-ac0    |                    |            |                   |       |                | agent                  |
| 1-f29db5d9d25    |                    |            |                   |       |                |                        |
| f                |                    |            |                   |       |                |                        |
| de622eb4-f320    | Linux bridge       | compute01  |                   | :-)   | True           | neutron-               |
| -466f-b2ae-      | agent              |            |                   |       |                | linuxbridge-           |
| b845b598f3b2     |                    |            |                   |       |                | agent                  |
+------------------+--------------------+------------+-------------------+-------+----------------+------------------------+
```

图8-10 列出公共网络服务代理

在正确配置的情况下，输出结果应该包括控制节点上的三个代理和每个计算节点上的一个代理。

b. 私有网络（Self-service networks），如图 8-11 所示。

```
[root@controller ~]# neutron agent-list
```

id	agent_type	host	availability_zone	alive	admin_state_up	binary
18c079b6-41f4-472d-8b3a-4ab13ee0e20b	Linux bridge agent	compute01		:-)	True	neutron-linuxbridge-agent
25ad08c6-d7ef-4af4-be6c-224ddd5b4560	Metadata agent	controller		:-)	True	neutron-metadata-agent
865a5980-6b80-40a6-9fdb-cdc7d8ea7f11	Linux bridge agent	controller		:-)	True	neutron-linuxbridge-agent
bbea0b81-e0a4-4440-b88f-d6d51d257102	DHCP agent	controller	nova	:-)	True	neutron-dhcp-agent
dc301d33-142d-498c-b073-b3691e4754b7	L3 agent	controller	nova	:-)	True	neutron-l3-agent

图 8-11　列出私有网络服务代理

在正确配置的情况下，输出结果应该包括控制节点上的四个代理和每个计算节点上的一个代理。

第 9 章 仪表板（Horzion）服务

9.1 服务概述

Horizon（Dashboard）是一个 Web 接口，云平台管理员以及用户可以通过它管理不同的 Openstack 资源以及服务。

9.2 安装和配置（控制节点）

9.2.1 安装软件包

安装 dashboard 软件包：
yum -y install openstack-dashboard

9.2.2 配置 Dashboard

编辑 /etc/openstack-dashboard/local_settings 文件：

由于默认配置文件在各发行版本中可能不同，因此，在进行修改的同时可能需要添加部分选项。另外，在配置片段中的省略号（...）表示默认的配置选项，应该保留。
gedit /etc/openstack-dashboard/local_settings

a. 允许所有主机访问仪表板：

ALLOWED_HOSTS = ['*',]

b. 配置 API 版本：

OPENSTACK_API_VERSIONS = {
 "identity": 3,
 "image": 2,
 "volume": 2,
}

c. 启用对域的支持：

OPENSTACK_KEYSTONE_MULTIDOMAIN_SUPPORT = True

d. 通过仪表板创建用户时的默认域配置为 default：

OPENSTACK_KEYSTONE_DEFAULT_DOMAIN = "default"

e. 配置 memcached 会话存储服务：

```
CACHES = {
    'default': {
        'BACKEND': 'django.core.cache.backends.memcached.MemcachedCache',
        'LOCATION': 'controller：11211',
    }
}
```

将其他的会话存储服务配置注释。

f. 在控制节点上配置仪表板以使用 OpenStack 服务：

OPENSTACK_HOST = "controller"

g. 启用第 3 版认证 API：

OPENSTACK_KEYSTONE_URL = "http://%s:5000/v3" % OPENSTACK_HOST

h. 通过仪表板创建的用户默认角色配置为 user：

OPENSTACK_KEYSTONE_DEFAULT_ROLE = "user"

i. 如果控制节点没有采用私有网络（Self-service networks）类型，则需禁用 Layer-3 网络服务的支持：

如果使用公共网络（Provider networks），则需修改此项的配置。

```
OPENSTACK_NEUTRON_NETWORK = {
    ...
    'enable_router': False,
    'enable_quotas': False,
    'enable_distributed_router': False,
    'enable_ha_router': False,
    'enable_lb': False,
    'enable_firewall': False,
    'enable_vpn': False,
    'enable_fip_topology_check': False,
}
```

j. 可以选择性地配置时区：

TIME_ZONE = " Asia/Shanghai"

9.2.3 完成安装

重启 Web 服务器以及会话存储服务：
systemctl restart httpd.service memcached.service

9.3 验证操作（控制节点）

使用 admin 或者 demo 用户凭证和 default 域凭证访问 OpenStack 进行验证。
1）打开菜单"应用程序" > "互联网" > "火狐浏览器"，如图 9-1 所示。

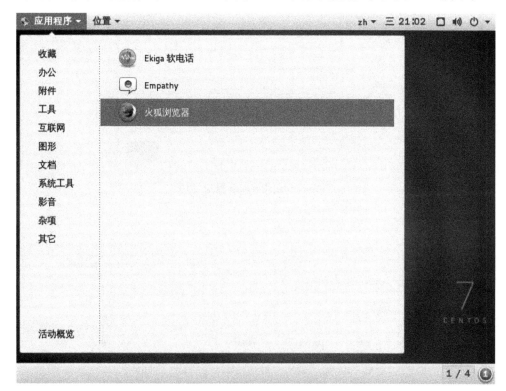

图 9-1 打开火狐浏览器

2）在 Web 浏览器中访问仪表板，地址为 http://controller/dashboard
可以使用 default 域的 admin 或 demo 用户凭证进行登录，如图 9-2 及图 9-3 所示。

图 9-2 使用 admin 登录（1）

图 9-3 使用 admin 登录（2）

如果仪表板界面显示的是英文界面，可以通过以下步骤设置语言环境。

a. 英文界面如图 9-4 所示，先单击右上角用户名，然后选择 Settings。

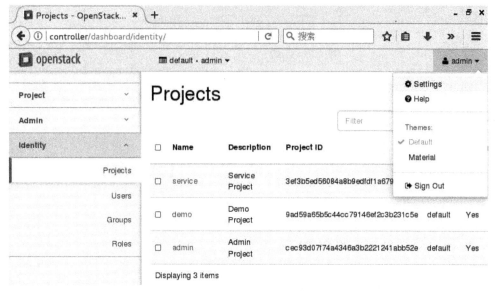

图 9-4　英文仪表板界面

b. 先在 User Settings 页面的 Language 列表中选择简体中文（zh-cn），然后单击 Save，如图 9-5 所示。

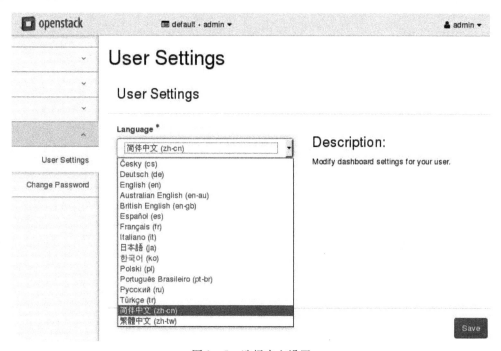

图 9-5　选择中文设置

c. 仪表板界面将显示为中文，如图 9-6 所示。

图 9-6 中文仪表板界面

第 10 章 块存储（Cinder）服务

10.1 服务概述

块存储服务为虚拟机实例提供块存储。存储的分配和消耗是由块存储驱动器或者多后端配置的驱动器决定的。还有很多驱动程序可用，例如 NAS/SAN、NFS、ISCSI 和 Ceph 等。

块存储是将磁盘整个映射给主机使用，提供逻辑卷功能。通常情况下，块存储服务 API 和调度器服务运行在控制节点上。在使用的驱动决定下，卷服务器可以运行在控制节点、计算节点或单独的块存储节点上。

OpenStack 块存储服务（Cinder）为虚拟机添加持久的存储，块存储提供一个基础设施，为了管理卷，以及和 OpenStack 计算服务交互，为实例提供卷。此服务也会激活管理卷的快照和卷类型的功能。

Cinder 块存储服务的核心功能是对卷的管理，允许对卷及其类型、快照进行处理。它并没有实现对块设备的管理和实际服务（提供逻辑卷），而是通过后端的统一存储接口支持不同块设备厂商的块存储服务，实现其驱动支持并与 OpenStack 进行整合。

块存储服务通常包含下列组件：

1. cinder-api

cinder-api 用于接受 API 请求，并将其转发到 cinder-volume 执行。

2. cinder-volume

cinder-volume 服务可以与块存储服务（例如 cinder-scheduler）进程进行直接交互，它也可以通过一个消息队列与存储服务进程进行交互。

cinder-volume 服务以响应送到块存储服务的读写请求的方式来维持状态。它也可以和多种存储提供者在驱动架构下进行交互。

3. cinder-scheduler daemon

它可以选择最优存储，提供节点来创建卷。

4. cinder-backup daemon

cinder-backup 服务向一个备份存储提供者提供任何种类备份卷，就像 cinder-volume 服务一样，它与多种存储提供者在驱动架构下进行交互。

5. 消息队列

在块存储的进程之间路由信息。

10.2 安装前准备（控制节点）

安装和配置块存储服务之前，必须创建一个数据库、服务凭证和 API 端点。

10.2.1 创建数据库

1）用数据库连接客户端以 root 用户连接到数据库服务器
mysql -u root -p'SQL_PASS'

2）创建 cinder 数据库
CREATE DATABASE cinder;

3）对 cinder 数据库授予恰当的权限
本书示例中将 cinder 数据库密码设置为 CINDER_DBPASS。
GRANT ALL PRIVILEGES ON cinder.* TO 'cinder'@'localhost' \
　　IDENTIFIED BY 'CINDER_DBPASS';

GRANT ALL PRIVILEGES ON cinder.* TO 'cinder'@'%' \
　　IDENTIFIED BY 'CINDER_DBPASS';

4）退出数据库客户端
exit

10.2.2 创建用户、服务实体和 API 端点

1. 获得 admin 凭证

获得 admin 凭证来获取只有管理员才能执行的命令的访问权限：
./root/admin-openrc

2. 创建 cinder 用户

1）创建 cinder 用户
openstack user create --domain default \
　--password CINDER_PASS cinder

本书示例中将 cinder 用户密码设置为 CINDER_PASS，如图 10-1 所示。

```
[root@controller ~]# openstack user create --domain default \
> --password CINDER_PASS cinder
+-----------+----------------------------------+
| Field     | Value                            |
+-----------+----------------------------------+
| domain_id | 1224d01d12d5451191c2f0aeefa6ad6a |
| enabled   | True                             |
| id        | e19a8f81cdcc44d6bae860f1f0ec43e3 |
| name      | cinder                           |
+-----------+----------------------------------+
```

图 10-1　创建 cinder 用户

2) 添加 admin 角色到 cinder 用户和 service 项目上
openstack role add--project service--user cinder admin
如图 10-2 所示。

```
[root@controller ~]# openstack role add --project service --user cinder admin
[root@controller ~]#
```

图 10-2　添加 admin 角色到 cinder 用户和 service 项目上

3. 创建服务实体和身份认证服务

1) 创建 cinder 和 cinderv2 服务实体

块存储服务要求两个服务实体，如图 10-3 及图 10-4 所示。

openstack service create--name cinder \
　--description "OpenStack Block Storage" volume

```
[root@controller ~]# openstack service create --name cinder \
> --description "OpenStack Block Storage" volume
+-------------+----------------------------------+
| Field       | Value                            |
+-------------+----------------------------------+
| description | OpenStack Block Storage          |
| enabled     | True                             |
| id          | 97f107b9379a4d10ab1eebdfb7af2fcd |
| name        | cinder                           |
| type        | volume                           |
+-------------+----------------------------------+
```

图 10-3　创建 cinder 服务实体

openstack service create--name cinderv2 \
　--description "OpenStack Block Storage" volumev2

```
[root@controller ~]# openstack service create --name cinderv2 \
> --description "OpenStack Block Storage" volumev2
+-------------+----------------------------------+
| Field       | Value                            |
+-------------+----------------------------------+
| description | OpenStack Block Storage          |
| enabled     | True                             |
| id          | bd5e0be7123444df8dca1f6729e5c798 |
| name        | cinderv2                         |
| type        | volumev2                         |
+-------------+----------------------------------+
```

图 10-4　创建 cinder 2 服务实体

4. 创建 API 端点

创建块存储服务 API 端点：

块存储服务每个服务实体都需要端点，分别如图 10-5 至图 10-10 所示。

\# openstack endpoint create--region RegionOne \

 volume public http：//controller：8776/v1/% \ (tenant_id \) s

```
[root@controller ~]# openstack endpoint create --region RegionOne \
> volume public http://controller:8776/v1/%\(tenant_id\)s
+--------------+----------------------------------------+
| Field        | Value                                  |
+--------------+----------------------------------------+
| enabled      | True                                   |
| id           | cb6575f17fc849ffbf80c41f05bde3e5       |
| interface    | public                                 |
| region       | RegionOne                              |
| region_id    | RegionOne                              |
| service_id   | 97f107b9379a4d10ab1eebdfb7af2fcd       |
| service_name | cinder                                 |
| service_type | volume                                 |
| url          | http://controller:8776/v1/%(tenant_id)s|
+--------------+----------------------------------------+
```

图 10-5　创建 cinder API 端点（public）

\# openstack endpoint create--region RegionOne \

 volume internal http：//controller：8776/v1/% \ (tenant_id \) s

```
[root@controller ~]# openstack endpoint create --region RegionOne \
> volume internal http://controller:8776/v1/%\(tenant_id\)s
+--------------+----------------------------------------+
| Field        | Value                                  |
+--------------+----------------------------------------+
| enabled      | True                                   |
| id           | 5c3fb8894d6b48448067df19e0ccaf59       |
| interface    | internal                               |
| region       | RegionOne                              |
| region_id    | RegionOne                              |
| service_id   | 97f107b9379a4d10ab1eebdfb7af2fcd       |
| service_name | cinder                                 |
| service_type | volume                                 |
| url          | http://controller:8776/v1/%(tenant_id)s|
+--------------+----------------------------------------+
```

图 10-6　创建 cinder API 端点（internal）

\# openstack endpoint create--region RegionOne \

 volume admin http：//controller：8776/v1/% \ (tenant_id \) s

```
[root@controller ~]# openstack endpoint create --region RegionOne \
> volume admin http://controller:8776/v1/%\(tenant_id\)s
+--------------+-------------------------------------------+
| Field        | Value                                     |
+--------------+-------------------------------------------+
| enabled      | True                                      |
| id           | 3d7e66401cad483bbcb5f2c5e56042eb          |
| interface    | admin                                     |
| region       | RegionOne                                 |
| region_id    | RegionOne                                 |
| service_id   | 97f107b9379a4d10ab1eebdfb7af2fcd          |
| service_name | cinder                                    |
| service_type | volume                                    |
| url          | http://controller:8776/v1/%(tenant_id)s   |
+--------------+-------------------------------------------+
```

图 10-7 创建 cinder API 端点（admin）

openstack endpoint create--region RegionOne \

volumev2 public http：//controller：8776/v2/% \ (tenant_id \) s

```
[root@controller ~]# openstack endpoint create --region RegionOne \
> volumev2 public http://controller:8776/v2/%\(tenant_id\)s
+--------------+-------------------------------------------+
| Field        | Value                                     |
+--------------+-------------------------------------------+
| enabled      | True                                      |
| id           | 47d1464ec5b64b1b8448bb20c8560437          |
| interface    | public                                    |
| region       | RegionOne                                 |
| region_id    | RegionOne                                 |
| service_id   | bd5e0be7123444df8dca1f6729e5c798          |
| service_name | cinderv2                                  |
| service_type | volumev2                                  |
| url          | http://controller:8776/v2/%(tenant_id)s   |
+--------------+-------------------------------------------+
```

图 10-8 创建 cinder 2 API 端点（public）

openstack endpoint create--region RegionOne \

volumev2 internal http：//controller：8776/v2/% \ (tenant_id \) s

```
[root@controller ~]# openstack endpoint create --region RegionOne \
> volumev2 internal http://controller:8776/v2/%\(tenant_id\)s
+--------------+-------------------------------------------+
| Field        | Value                                     |
+--------------+-------------------------------------------+
| enabled      | True                                      |
| id           | fa71b37070414ed8b9388d8e7c9a6262          |
| interface    | internal                                  |
| region       | RegionOne                                 |
| region_id    | RegionOne                                 |
| service_id   | bd5e0be7123444df8dca1f6729e5c798          |
| service_name | cinderv2                                  |
| service_type | volumev2                                  |
| url          | http://controller:8776/v2/%(tenant_id)s   |
+--------------+-------------------------------------------+
```

图 10-9 创建 cinder 2 API 端点（internal）

openstack endpoint create--region RegionOne \
volumev2 admin http：//controller：8776/v2/% \ (tenant_id \) s

```
[root@controller ~]# openstack endpoint create --region RegionOne \
> volumev2 admin http://controller:8776/v2/%\(tenant_id\)s
+--------------+-------------------------------------------+
| Field        | Value                                     |
+--------------+-------------------------------------------+
| enabled      | True                                      |
| id           | 601862d6610841dba73b38f7aef81e6c          |
| interface    | admin                                     |
| region       | RegionOne                                 |
| region_id    | RegionOne                                 |
| service_id   | bd5e0be7123444df8dca1f6729e5c798          |
| service_name | cinderv2                                  |
| service_type | volumev2                                  |
| url          | http://controller:8776/v2/%(tenant_id)s   |
+--------------+-------------------------------------------+
```

图 10 - 10　创建 cinder 2 API 端点（admin）

10.3　安装和配置（控制节点）

10.3.1　安装软件包

安装 cinder 软件包：
yum-y install openstack-cinder

10.3.2　配置 Cinder

编辑 /etc/cinder/cinder.conf 文件：

由于默认配置文件在各发行版本中可能不同，因此，在进行修改的同时可能需要添加部分选项。另外，在配置片段中的省略号（...）表示默认的配置选项，应该保留。
gedit /etc/cinder/cinder.conf

a. 在 [database] 部分，配置数据库访问：

[database]
...
connection = mysql + pymysql：//cinder：CINDER_DBPASS@controller/cinder

b. 在 [DEFAULT] 和 [oslo_messaging_rabbit] 部分，配置 RabbitMQ 消息队列访问：

```
[DEFAULT]
...
rpc_backend = rabbit

[oslo_messaging_rabbit]
...
rabbit_host = controller
rabbit_userid = openstack
rabbit_password = RABBIT_PASS
```

c. 在[DEFAULT]和[keystone_authtoken]部分，配置认证服务访问：

```
[DEFAULT]
...
auth_strategy = keystone

[keystone_authtoken]
...
auth_uri = http://controller:5000
auth_url = http://controller:35357
memcached_servers = controller:11211
auth_type = password
project_domain_name = default
user_domain_name = default
project_name = service
username = cinder
password = CINDER_PASS
```

在[keystone_authtoken]中注释或者删除其他选项。

d. 在[DEFAULT]部分，配置my_ip来使用控制节点的管理接口的IP地址：

```
[DEFAULT]
...
my_ip = MANAGEMENT_INTERFACE_IP_ADDRESS
```

将MANAGEMENT_INTERFACE_IP_ADDRESS替换为控制节点的管理网络IP地址，本书示例中将其替换为192.168.1.11。

e. 在 [oslo_concurrency] 部分，配置锁路径：

[oslo_concurrency]
...
lock_path = /var/lib/cinder/tmp

10.3.3 为计算服务配置块存储服务

编辑 /etc/nova/nova.conf 文件：

由于默认配置文件在各发行版本中可能不同，因此，在进行修改的同时可能需要添加部分选项。另外，在配置片段中的省略号（...）表示默认的配置选项，应该保留。

gedit /etc/nova/nova.conf

在 [cinder] 部分，配置访问参数：

[cinder]
...
os_region_name = RegionOne

10.3.4 写入数据库

写入块存储服务数据库：

su-s /bin/sh-c "cinder-manage db sync" cinder

如图 10-11 所示。

```
[root@controller ~]# su -s /bin/sh -c "cinder-manage db sync" cinder
Option "logdir" from group "DEFAULT" is deprecated. Use option "log-dir" from group "DEFAULT".
2017-03-27 15:39:43.908 4442 WARNING py.warnings [-] /usr/lib/python2.7/site-packages/oslo_db/sqlal
chemy/enginefacade.py:241: NotSupportedWarning: Configuration option(s) ['use_tpool'] not supported
 exception.NotSupportedWarning

2017-03-27 15:39:44.016 4442 INFO migrate.versioning.api [-] 0 -> 1...
2017-03-27 15:39:44.307 4442 INFO migrate.versioning.api [-] done
2017-03-27 15:39:44.308 4442 INFO migrate.versioning.api [-] 1 -> 2...
2017-03-27 15:39:44.414 4442 INFO migrate.versioning.api [-] done
```

图 10-11 写入块存储服务数据库

忽略输出中任何不推荐使用的信息。

10.3.5 完成安装

1) 重启计算 API 服务

systemctl restart openstack-nova-api.service

2) 启动块存储服务，并将其配置为随系统启动

systemctl enable openstack-cinder-api.service \

openstack-cinder-scheduler. service
systemctl start openstack-cinder-api. service \
openstack-cinder-scheduler. service

10.4 安装和配置（块存储节点）

本节示例中使用了一个本地硬盘（/dev/sdb）作为块存储设备，将在该设备上使用 LVM（Logical Volume Manager）提供逻辑卷，并通过 iSCSI 协议提供给实例使用。

可以通过本节的配置方法，添加额外的块存储节点来增加块存储的环境规模。

10.4.1 准备 LVM 块存储卷

1）安装 lvm 软件包
yum-y install lvm2
2）启动 LVM 的 metadata 服务，并将其配置为随系统启动
systemctl enable lvm2-lvmetad. service
systemctl start lvm2-lvmetad. service
3）创建 LVM 物理卷 /dev/sdb
pvcreate /dev/sdb
如图 10-12 所示。

```
[root@blockstorage01 ~]# pvcreate /dev/sdb
  Physical volume "/dev/sdb" successfully created.
```

图 10-12　创建 LVM 物理卷/dev/sdb

4）创建 LVM 卷组 cinder-volumes
vgcreate cinder-volumes /dev/sdb
如图 10-13 所示。

```
[root@blockstorage01 ~]# vgcreate cinder-volumes /dev/sdb
  Volume group "cinder-volumes" successfully created
```

图 10-13　创建 LVM 卷组 cinder-volumes

块存储服务会在这个 cinder-volumes 卷组中创建逻辑卷。

5）重新配置 LVM

虽然块存储卷只有虚拟机实例才能访问，但是块存储节点底层的操作系统可以管理这些设备并将其与卷关联。默认情况下，LVM 卷扫描工具会扫描/dev 目录下的所有设备，查找包含虚拟机实例访问的块存储卷。如果块存储卷被虚拟机实例用作 LVM，当 LVM 卷扫描工具检测到这些卷时会尝试缓存它们，则可能会在底层操作系统和块存储卷上产生各

种问题。因此必须重新配置 LVM，让它只扫描包含 cinder-volumes 卷组的设备。

6）编辑 /etc/lvm/lvm.conf 文件

由于默认配置文件在各发行版本中可能不同，因此，在进行修改的同时可能需要添加部分选项。另外，在配置片段中的省略号（...）表示默认的配置选项，应该保留。

gedit /etc/lvm/lvm.conf

在 devices 部分，添加一个过滤器：

每个过滤器组（filter）中的元素都以 a（accept）或 r（reject）开头，并包括一个设备名称的正则表达式规则，同时过滤器组必须以 r/.*/ 结束，过滤所有保留设备。可以使用 vgs-vvvv 命令来测试过滤器是否生效。

如果块存储节点关联到 Cinder 卷组中的设备使用了 LVM，则只需将其设备添加到过滤器中。例如，如果 /dev/sdb 设备使用了 LVM 并关联了 Cinder 卷组，则添加过滤器如下：

filter = ["a/sdb/", " r/.*/"]

如果块存储节点在操作系统磁盘上使用了 LVM，则还必须添加相关的设备到过滤器中。例如，如果 /dev/sda 设备包含操作系统，则添加过滤器如下：

filter = ["a/sda/", "a/sdb/", " r/.*/"]

类似地，如果计算节点在操作系统磁盘上使用了 LVM，则也必须修改这些节点上 /etc/lvm/lvm.conf 文件中的过滤器，将操作系统磁盘包含到过滤器中。例如，如果 /dev/sda 设备包含操作系统，则添加过滤器如下：

filter = ["a/sda/", " r/.*/"]

在本书示例中，CentOS 操作系统安装到了 /dev/sda 磁盘中并使用了 LVM。同时，/dev/sdb 磁盘将用于块存储，因此，在 devices 部分，添加如下一个过滤器：

devices {
...
filter = ["a/sda/", "a/sdb/", " r/.*/"]

10.4.2 安装软件包

安装 cinder 软件包：

yum -y install openstack-cinder targetcli python-keystone

10.4.3 配置 Cinder

编辑 /etc/cinder/cinder.conf 文件：

由于默认配置文件在各发行版本中可能不同,因此,在进行修改的同时可能需要添加部分选项。另外,在配置片段中的省略号(...)表示默认的配置选项,应该保留。

gedit /etc/cinder/cinder.conf

a. 在[database]部分,配置数据库访问:

[database]
...
connection = mysql + pymysql://cinder:CINDER_DBPASS@controller/cinder

b. 在[DEFAULT]和[oslo_messaging_rabbit]部分,配置RabbitMQ消息队列访问:

[DEFAULT]
...
rpc_backend = rabbit

[oslo_messaging_rabbit]
...
rabbit_host = controller
rabbit_userid = openstack
rabbit_password = RABBIT_PASS

c. 在[DEFAULT]和[keystone_authtoken]部分,配置认证服务访问:

[DEFAULT]
...
auth_strategy = keystone

[keystone_authtoken]
...
auth_uri = http://controller:5000
auth_url = http://controller:35357
memcached_servers = controller:11211
auth_type = password
project_domain_name = default
user_domain_name = default
project_name = service
username = cinder
password = CINDER_PASS

在［keystone_authtoken］中注释或者删除其他选项。

d. 在［DEFAULT］部分，配置 my_ip 来使用块存储节点的管理接口 IP 地址：

［DEFAULT］
...
my_ip = MANAGEMENT_INTERFACE_IP_ADDRESS

将 MANAGEMENT_INTERFACE_IP_ADDRESS 替换为块存储节点的管理网络 IP 地址，本书示例中将其替换为 192.168.1.41。

e. 在［DEFAULT］部分，启用 LVM 后端：

［DEFAULT］
...
enabled_backends = lvm

后端名字是任意的。比如，本书示例中使用驱动的名字作为后端的名字。

f. 在［lvm］部分，配置 LVM 后端驱动、卷组、iSCSI 协议和 iSCSI 服务：

由于初始配置文件中并没有［lvm］部分，因此可以在文件尾部添加此部分内容。

［lvm］
volume_driver = cinder.volume.drivers.lvm.LVMVolumeDriver
volume_group = cinder-volumes
iscsi_protocol = iscsi
iscsi_helper = lioadm

g. 在［DEFAULT］部分，配置镜像服务 API 的位置：

［DEFAULT］
...
glance_api_servers = http：//controller：9292

h. 在［oslo_concurrency］部分，配置锁路径：

［oslo_concurrency］
...
lock_path = /var/lib/cinder/tmp

10.4.4 完成安装

启动块存储服务，并将其配置为随系统启动：

systemctl enable openstack-cinder-volume.service target.service
systemctl start openstack-cinder-volume.service target.service

10.5 验证操作（控制节点）

1）获得 admin 凭证来获取只有管理员才能执行的命令的访问权限
#./root/admin-openrc
2）列出服务组件，以验证每个进程是否都成功启动
cinder service-list
如图 10-14 所示。

```
[root@controller ~]# cinder service-list
+------------------+--------------------+------+---------+-------+----------------------------+-----------------+
|      Binary      |        Host        | Zone | Status  | State |         Updated_at         | Disabled Reason |
+------------------+--------------------+------+---------+-------+----------------------------+-----------------+
| cinder-scheduler |     controller     | nova | enabled |   up  | 2017-03-27T07:54:59.000000 |        -        |
|  cinder-volume   | blockstorage01@lvm | nova | enabled |   up  | 2017-03-27T07:54:19.000000 |        -        |
+------------------+--------------------+------+---------+-------+----------------------------+-----------------+
```

图 10-14 列出服务组件

在正确配置的情况下，输出结果应该包括控制节点上的一个代理和每个块存储节点上的一个代理。

如果块存储节点的状态显示为 down，且各项配置都正确时，则可以检查相关节点主机是否与控制节点的时间保持同步。

如图 10-15 所示，由于两个控制节点与块存储之间的时间误差太大，导致块存储节点中的服务异常。

```
[root@controller ~]# cinder service-list
+------------------+--------------------+------+---------+-------+----------------------------+-----------------+
|      Binary      |        Host        | Zone | Status  | State |         Updated_at         | Disabled Reason |
+------------------+--------------------+------+---------+-------+----------------------------+-----------------+
| cinder-scheduler |     controller     | nova | enabled |   up  | 2017-03-27T07:53:19.000000 |        -        |
|  cinder-volume   | blockstorage01@lvm | nova | enabled |  down | 2017-03-27T07:50:58.000000 |        -        |
+------------------+--------------------+------+---------+-------+----------------------------+-----------------+
```

图 10-15 时间误差太大导致服务异常

第 11 章　文件共享（Manila）服务

为简单起见，本书示例中将文件共享服务安装到块存储节点中，也可根据实际需要部署独立的文件共享节点。

11.1　服务概述

manila 服务用于协调共享文件系统或分布式文件系统。它既可以支持文件共享服务器或者多驱动器中的驱动器，也可以支持 NFS、CIFS、HDFS 协议的驱动器。

通常情况下，文件共享系统的 API 和调度服务运行在控制节点上，依赖于运行在控制节点、计算节点或者文件共享（块存储）节点上的共享服务驱动器。

共享文件服务为实例提供文件存储。共享文件服务为管理和配置共享文件提供基础设施。如果驱动支持快照，那么该服务也可管理共享快照的共享文件类型。

11.1.1　文件共享服务支持两种共享服务器管理支持模式

1. 没有驱动支持的文件共享服务器管理

在这种模式下，服务不需要任何和网络有关的部署。操作者只需确保实例和 NFS 服务器之间的连接。本选项使用需要包含 LVM 和 NFS 包以及一个额外的命名为 manila-share 的 LVM 卷组的 LVM 驱动器。

2. 有驱动支持的文件共享服务器管理

在这种模式下，服务需要计算（nova）、网络（neutron）和块存储（cinder）服务来管理共享服务器。这部分信息用于创建共享服务器，就像创建共享网络一样。本选项使用支持共享服务处理的 generic 驱动器，并且需要一个连接到路由的私有网络（Self-service networks）。

11.1.2　共享文件系统服务包含的组件

1. manila-api

一个 WSGI 应用，通过共享文件系统服务来认证和路由请求，同时也支持 OpenStack 的 APIs。

2. manila-data

一个单一服务的功能组件，包括了接受请求、处理数据操作例如拷贝、迁移、备份，

3. manila-scheduler

把请求调度和路由到合适的共享服务端点。调度器用可配置的过滤器和权重计算器来路由请求。过滤调度器是默认的模式，它可以对很多对象进行过滤，比如：容量、可用区和共享类型，当然也可以自定义一些自己的过滤器。

4. manila-share

这个共享节点支持两种模式，即带有和不带有共享服务器处理两种模式。这两种模式都依赖于驱动器的支持。

5. 消息队列

在共享文件系统进程之间的路由信息。

11.2 安装前准备（控制节点）

安装和配置文件共享服务之前，必须创建一个数据库、服务凭证和 API 端点。

11.2.1 创建数据库

1）用数据库连接客户端以 root 用户连接到数据库服务器
mysql-u root-p'SQL_PASS'
2）创建 manila 数据库
CREATE DATABASE manila;
3）对 manila 数据库授予恰当的权限
本书示例中将 manila 数据库密码设置为 MANILA_DBPASS。
GRANT ALL PRIVILEGES ON manila.* TO 'manila'@'localhost' \
  IDENTIFIED BY 'MANILA_DBPASS';
GRANT ALL PRIVILEGES ON manila.* TO 'manila'@'%' \
  IDENTIFIED BY 'MANILA_DBPASS';
4）退出数据库客户端
exit

11.2.2 创建用户、服务实体和 API 端点

1. 获得 admin 凭证

获得 admin 凭证来获取只有管理员才能执行的命令的访问权限：
./root/admin-openrc

2. 创建 manila 用户

1）创建 manila 用户

openstack user create--domain default \
　　--password MANILA_PASS manila

本书示例中将 manila 用户密码设置为 MANILA_PASS，如图 11-1 所示。

```
[root@controller ~]# openstack user create --domain default \
> --password MANILA_PASS manila
+-----------+----------------------------------+
| Field     | Value                            |
+-----------+----------------------------------+
| domain_id | 1224d01d12d5451191c2f0aeefa6ad6a |
| enabled   | True                             |
| id        | 5fca620121ac433a89c526a10a210b93 |
| name      | manila                           |
+-----------+----------------------------------+
```

图 11-1　创建 manila 用户

2）添加 admin 角色到 manila 用户和 service 项目上

openstack role add--project service--user manila admin

如图 11-2 所示。

```
[root@controller ~]# openstack role add --project service --user manila admin
[root@controller ~]#
```

图 11-2　添加 admin 角色到 manila 用户和 servile 项目上

3. 创建服务实体和身份认证服务

创建 manila 和 manilav2 服务实体：

文件共享服务要求两个服务实体，分别如图 11-3 和图 11-4 所示。

openstack service create--name manila \
　　--description "OpenStack Shared File Systems" share

```
[root@controller ~]# openstack service create --name manila \
> --description "OpenStack Shared File Systems" share
+-------------+----------------------------------+
| Field       | Value                            |
+-------------+----------------------------------+
| description | OpenStack Shared File Systems    |
| enabled     | True                             |
| id          | c3ccbec63e574fdf8608db7921c1b856 |
| name        | manila                           |
| type        | share                            |
+-------------+----------------------------------+
```

图 11-3　创建 manila 服务实体

openstack service create--name manilav2 \
　　--description "OpenStack Shared File Systems" sharev2

图 11-4 创建 manilav 2 服务实体

4. 创建 API 端点

创建文件共享服务 API 端点：

文件共享服务每个服务实体都需要端点，分别如图 11-5 至图 11-10 所示。

openstack endpoint create--region RegionOne \
　　share public http：//controller：8786/v1/% \ （tenant_id\）s

```
[root@controller ~]# openstack endpoint create --region RegionOne \
> share public http://controller:8786/v1/%\(tenant_id\)s
+--------------+-------------------------------------------+
| Field        | Value                                     |
+--------------+-------------------------------------------+
| enabled      | True                                      |
| id           | 9e71dc4fe769474ab6b9e2162222c6b3          |
| interface    | public                                    |
| region       | RegionOne                                 |
| region_id    | RegionOne                                 |
| service_id   | c3ccbec63e574fdf8608db7921c1b856          |
| service_name | manila                                    |
| service_type | share                                     |
| url          | http://controller:8786/v1/%(tenant_id)s   |
+--------------+-------------------------------------------+
```

图 11-5 创建 manila API 端点（public）

openstack endpoint create--region RegionOne \
　　share internal http：//controller：8786/v1/% \ （tenant_id\）s

```
[root@controller ~]# openstack endpoint create --region RegionOne \
> share internal http://controller:8786/v1/%\(tenant_id\)s
+--------------+-------------------------------------------+
| Field        | Value                                     |
+--------------+-------------------------------------------+
| enabled      | True                                      |
| id           | 969a5b51e2b6416f8ea6d0fa60602e92          |
| interface    | internal                                  |
| region       | RegionOne                                 |
| region_id    | RegionOne                                 |
| service_id   | c3ccbec63e574fdf8608db7921c1b856          |
| service_name | manila                                    |
| service_type | share                                     |
| url          | http://controller:8786/v1/%(tenant_id)s   |
+--------------+-------------------------------------------+
```

图 11-6 创建 manila API 端点（internal）

openstack endpoint create--region RegionOne \

　　share admin http：//controller：8786/v1/% \ （tenant_id \）s

```
[root@controller ~]# openstack endpoint create --region RegionOne \
> share admin http://controller:8786/v1/%\(tenant_id\)s
+--------------+-------------------------------------------+
| Field        | Value                                     |
+--------------+-------------------------------------------+
| enabled      | True                                      |
| id           | a5eadfe948aa45b48301da96b1fc740f          |
| interface    | admin                                     |
| region       | RegionOne                                 |
| region_id    | RegionOne                                 |
| service_id   | c3ccbec63e574fdf8608db7921c1b856          |
| service_name | manila                                    |
| service_type | share                                     |
| url          | http://controller:8786/v1/%(tenant_id)s   |
+--------------+-------------------------------------------+
```

图 11-7 创建 manila API 端点（admin）

openstack endpoint create--region RegionOne \

　　sharev2 public http：//controller：8786/v2/% \ （tenant_id \）s

```
[root@controller ~]# openstack endpoint create --region RegionOne \
> sharev2 public http://controller:8786/v2/%\(tenant_id\)s
+--------------+-------------------------------------------+
| Field        | Value                                     |
+--------------+-------------------------------------------+
| enabled      | True                                      |
| id           | bf9bf677bcb1403cad5da3649bc4c103          |
| interface    | public                                    |
| region       | RegionOne                                 |
| region_id    | RegionOne                                 |
| service_id   | 224de18754a14db8b5fed60fdea067b7          |
| service_name | manilav2                                  |
| service_type | sharev2                                   |
| url          | http://controller:8786/v2/%(tenant_id)s   |
+--------------+-------------------------------------------+
```

图 11-8 创建 manilave 2 API 端点（public）

```
# openstack endpoint create--region RegionOne \
    sharev2 internal http：//controller：8786/v2/% \ （tenant_id \）s
```

```
[root@controller ~]# openstack endpoint create --region RegionOne \
> sharev2 internal http://controller:8786/v2/%\(tenant_id\)s
+--------------+----------------------------------------------+
| Field        | Value                                        |
+--------------+----------------------------------------------+
| enabled      | True                                         |
| id           | 828e814c13854fd8af009206b31c7db1             |
| interface    | internal                                     |
| region       | RegionOne                                    |
| region_id    | RegionOne                                    |
| service_id   | 224de18754a14db8b5fed60fdea067b7             |
| service_name | manilav2                                     |
| service_type | sharev2                                      |
| url          | http://controller:8786/v2/%(tenant_id)s      |
+--------------+----------------------------------------------+
```

图 11-9　创建 manilav 2 API 端点（internal）

```
# openstack endpoint create--region RegionOne \
    sharev2 admin http：//controller：8786/v2/% \ （tenant_id \）s
```

```
[root@controller ~]# openstack endpoint create --region RegionOne \
> sharev2 admin http://controller:8786/v2/%\(tenant_id\)s
+--------------+----------------------------------------------+
| Field        | Value                                        |
+--------------+----------------------------------------------+
| enabled      | True                                         |
| id           | 34cff29308e54c4e8208f22198484305             |
| interface    | admin                                        |
| region       | RegionOne                                    |
| region_id    | RegionOne                                    |
| service_id   | 224de18754a14db8b5fed60fdea067b7             |
| service_name | manilav2                                     |
| service_type | sharev2                                      |
| url          | http://controller:8786/v2/%(tenant_id)s      |
+--------------+----------------------------------------------+
```

图 11-10　创建 manilave 2 API 端点（admin）

11.3　安装和配置（控制节点）

11.3.1　安装软件包

安装 manila 软件包：
yum-y install openstack-manila python-manilaclient

11.3.2　配置 Manila

编辑 /etc/manila/manila.conf 文件：

由于默认配置文件在各发行版本中可能不同,因此,在进行修改的同时可能需要添加部分选项。另外,在配置片段中的省略号(...)表示默认的配置选项,应该保留。

gedit /etc/manila/manila.conf

a. 在[database]部分,配置数据库访问:

[database]
...
connection = mysql + pymysql: //manila: MANILA_DBPASS@ controller/manila

b. 在[DEFAULT]和[oslo_messaging_rabbit]部分,配置RabbitMQ消息队列访问:

[DEFAULT]
...
rpc_backend = rabbit

[oslo_messaging_rabbit]
...
rabbit_host = controller
rabbit_userid = openstack
rabbit_password = RABBIT_PASS

c. 在[DEFAULT]和[keystone_authtoken]部分,配置认证服务访问:

[DEFAULT]
...
auth_strategy = keystone

[keystone_authtoken]
...
auth_uri = http: //controller: 5000
auth_url = http: //controller: 35357
memcached_servers = controller: 11211
auth_type = password
project_domain_name = default
user_domain_name = default
project_name = service
username = manila
password = MANILA_PASS

在[keystone_authtoken]中注释或者删除其他选项。

d. 在［DEFAULT］部分，配置文件共享类型：

［DEFAULT］
…
default_share_type = default_share_type
rootwrap_config = /etc/manila/rootwrap.conf

e. 在［DEFAULT］部分，配置 my_ip 来使用控制节点的管理接口的 IP 地址：

［DEFAULT］
…
my_ip = MANAGEMENT_INTERFACE_IP_ADDRESS

将 MANAGEMENT_INTERFACE_IP_ADDRESS 替换为控制节点的管理网络 IP 地址，本书示例中将其替换为 192.168.1.11。

f. 在［oslo_concurrency］部分，配置锁路径：

［oslo_concurrency］
…
lock_path = /var/lib/manila/tmp

11.3.3 写入数据库

写入文件共享服务数据库：

su-s /bin/sh-c "manila-manage db sync" manila

如图 11-11 所示。

```
[root@controller ~]# su -s /bin/sh -c "manila-manage db sync" manila
2017-03-27 16:13:18.994 4501 INFO alembic.runtime.migration [-] Context impl MySQLImpl.
2017-03-27 16:13:18.995 4501 INFO alembic.runtime.migration [-] Will assume non-transactional DDL.
2017-03-27 16:13:19.046 4501 INFO alembic.runtime.migration [-] Running upgrade  -> 162a3e673105, m
anila_init
2017-03-27 16:13:19.564 4501 INFO alembic.runtime.migration [-] Running upgrade 162a3e673105 -> 211
836bf835c, add access level
2017-03-27 16:13:19.599 4501 INFO alembic.runtime.migration [-] Running upgrade 211836bf835c -> 38e
632621e5a, change volume_type to share_type
```

图 11-11　写入文件共享服务数据库

忽略输出中任何不推荐使用的信息。

11.3.4 完成安装

启动文件共享服务，并将其配置为随系统启动：
systemctl enable openstack-manila-api.service \

openstack-manila-scheduler.service
systemctl start openstack-manila-api.service \
openstack-manila-scheduler.service

11.4 安装和配置(块存储节点)

11.4.1 安装软件包

安装 manila 软件包:
yum -y install openstack-manila openstack-manila-share

11.4.2 配置 Manila

编辑 /etc/manila/manila.conf 文件:

由于默认配置文件在各发行版本中可能不同,因此,在进行修改的同时可能需要添加部分选项。另外,在配置片段中的省略号(...)表示默认的配置选项,应该保留。

gedit /etc/manila/manila.conf

a. 在[database]部分,配置数据库访问:

[database]
...
connection = mysql+pymysql://manila:MANILA_DBPASS@controller/manila

b. 在[DEFAULT]和[oslo_messaging_rabbit]部分,配置 RabbitMQ 消息队列访问:

[DEFAULT]
...
rpc_backend = rabbit

[oslo_messaging_rabbit]
...
rabbit_host = controller
rabbit_userid = openstack
rabbit_password = RABBIT_PASS

c. 在[DEFAULT]部分,配置文件共享类型:

```
[DEFAULT]
...
default_share_type = default_share_type
rootwrap_config = /etc/manila/rootwrap.conf
```

d. 在［DEFAULT］和［keystone_authtoken］部分，配置认证服务访问：

```
[DEFAULT]
...
auth_strategy = keystone

[keystone_authtoken]
...
auth_uri = http://controller:5000
auth_url = http://controller:35357
memcached_servers = controller:11211
auth_type = password
project_domain_name = default
user_domain_name = default
project_name = service
username = manila
password = MANILA_PASS
```

在［keystone_authtoken］中注释或者删除其他选项。

e. 在［DEFAULT］部分，配置 my_ip 来使用块文件共享（块存储）节点的管理接口 IP 地址：

```
[DEFAULT]
...
my_ip = MANAGEMENT_INTERFACE_IP_ADDRESS
```

将 MANAGEMENT_INTERFACE_IP_ADDRESS 替换为块文件共享（块存储）节点的管理网络 IP 地址，本书示例中将其替换为 192.168.1.41。

f. 在［oslo_concurrency］部分，配置锁路径：

[oslo_concurrency]
...
lock_path =/var/lib/manila/tmp

11.4.3 配置共享服务器管理支持模式

为简单起见，本书示例中在块存储节点中配置文件共享服务。

11.4.3.1. 模式1：没有驱动支持的文件共享服务器管理

LVM 驱动器需要一个单独的空的本地存储设备来避免与块存储服务产生冲突。本节示例中将使用 /dev/sdc 磁盘，作为文件共享服务的本地存储。

1. 安装软件包

1）安装 LVM 和 NFS 软件包

\# yum-y install lvm2 nfs-utils nfs4-acl-tools portmap

2）启动 LVM 的 metadata 服务，并将其配置为随系统启动

\# systemctl enable lvm2-lvmetad.service

\# systemctl start lvm2-lvmetad.service

2. 重新配置 LVM

虽然文件共享服务只有虚拟机实例才能访问，但是文件共享节点底层的操作系统也可以管理这些设备并将其与卷关联。默认情况下，LVM 卷扫描工具会扫描 /dev 目录下的所有设备，查找包含虚拟机实例访问的文件共享卷。如果文件共享卷被虚拟机实例用作 LVM，当 LVM 卷扫描工具检测到这些卷时会尝试缓存它们，则可能会在底层操作系统和块存储卷上产生各种问题。因此必须重新配置 LVM，让它只扫描包含 manila-volumes 卷组的设备。

1）编辑 /etc/lvm/lvm.conf 文件

由于默认配置文件在各发行版本中可能不同，因此，在进行修改的同时可能需要添加部分选项。另外，在配置片段中的省略号（...）表示默认的配置选项，应该保留。

\# gedit /etc/lvm/lvm.conf

在 devices 部分，添加一个过滤器：

由于每个过滤器组（filter）中的元素都以 a（accept）或 r（reject）开头，并包括一个设备名称的正则表达式规则，同时过滤器组必须以 r/.*/ 结束，过滤所有保留设备。因此，可以使用 vgs-vvvv 命令来测试过滤器是否生效。

如果块存储节点只有关联到 Manila 卷组中的设备使用了 LVM，则只需将其设备添加到过滤器中。例如，如果 /dev/sdc 设备使用了 LVM 并关联了 Manila 卷组，则添加过滤器如下：

```
filter = [ "a/sdc/"," r/.*/"]
```

如果块存储节点在操作系统磁盘上使用了 LVM，则还必须添加相关的设备到过滤器中。例如，如果 /dev/sda 设备包含操作系统，则添加过滤器如下：

```
filter = [ "a/sda/","a/sdc/"," r/.*/"]
```

类似地，如果计算节点在操作系统磁盘上使用了 LVM，则也必须修改这些节点上 /etc/lvm/lvm.conf 文件中的过滤器，将操作系统磁盘包含到过滤器中。例如，如果 /dev/sda 设备包含操作系统，则添加过滤器如下：

```
filter = [ "a/sdc/"," r/.*/"]
```

在本书示例中，由于 CentOS 操作系统安装到了 /dev/sda 磁盘中并使用了 LVM。同时，/dev/sdb 磁盘用于块存储，/dev/sdc 磁盘将用于文件共享，因此，在 devices 部分，添加如下一个过滤器：

```
devices {
...
filter = [ "a/sda/","a/sdb/","a/sdc/"," r/.*/"]
```

2）创建 LVM 物理卷 /dev/sdc

pvcreate /dev/sdc

如图 11-12 所示。

```
[root@blockstorage01 ~]# pvcreate /dev/sdc
  Physical volume "/dev/sdc" successfully created.
[root@blockstorage01 ~]#
```

图 11-12　创建 LVM 物理卷

3）创建 LVM 卷组 manila-volumes

vgcreate manila-volumes /dev/sdc

如图 11-13 所示。

```
[root@blockstorage01 ~]# vgcreate manila-volumes /dev/sdc
  Volume group "manila-volumes" successfully created
```

图 11-13　创建 LVM 卷组

文件共享服务会在这个 manila-volumes 卷组中创建逻辑卷。

3. 配置 Manila

编辑 /etc/manila/manila.conf 文件：

由于默认配置文件在各发行版本中可能不同，因此，在进行修改的同时可能需要添加部分选项。另外，在配置片段中的省略号（...）表示默认的配置选项，应该保留。

gedit /etc/manila/manila.conf

a. 在［DEFAULT］部分，启用 LVM 驱动和 NFS/CIFS 协议：

```
[DEFAULT]
...
enabled_share_backends = lvm
enabled_share_protocols = NFS,CIFS
```

b. 在［lvm］部分，配置 LVM 驱动、卷组、配置 lvm_share_export_ip 来使用文件共享节点管理接口的 IP 地址。

由于初始配置文件中并没有［lvm］部分，因此可以在文件尾部添加此部分内容。

```
[lvm]
share_backend_name = LVM
share_driver = manila.share.drivers.lvm.LVMShareDriver
driver_handles_share_servers = False
lvm_share_volume_group = manila-volumes
lvm_share_export_ip = MANAGEMENT_INTERFACE_IP_ADDRESS
```

将 MANAGEMENT_INTERFACE_IP_ADDRESS 替换为块文件共享（块存储）节点的管理网络 IP 地址，本书示例中将其替换为 192.168.1.41。

11.4.3.2 模式2：有驱动支持的文件共享服务器管理

在进行下一步之前，请验证计算、网络和块存储服务是否正常，并且需要一个连接到路由的私有网络（Self-service networks）的支持，同样也需要在块文件共享（块存储）节点上安装一些网络服务组件。

1. 安装软件包

安装 neutron 网络服务组件：

yum -y install openstack-neutron openstack-neutron-linuxbridge ebtables

2. 配置 Manila

编辑 /etc/manila/manila.conf 文件：

由于默认配置文件在各发行版本中可能不同，因此，在进行修改的同时可能需要添加部分选项。另外，在配置片段中的省略号（...）表示默认的配置选项，应该保留。

gedit /etc/manila/manila.conf

a. 在［DEFAULT］部分，启用 generic 驱动和 NFS/CIFS 协议：

```
[DEFAULT]
...
enabled_share_backends = generic
enabled_share_protocols = NFS,CIFS
```

b. 在[neutron]、[nova]、[cinder]部分，允许对这些服务的认证：

```
[neutron]
...
url = http://controller:9696
region_name = RegionOne
memcached_servers = controller:11211
auth_uri = http://controller:5000
auth_url = http://controller:35357
auth_type = password
project_domain_name = default
project_name = service
user_domain_name = default
username = neutron
password = NEUTRON_PASS

[nova]
...
memcached_servers = controller:11211
region_name = RegionOne
auth_uri = http://controller:5000
auth_url = http://controller:35357
auth_type = password
project_domain_name = default
project_name = service
user_domain_name = default
username = nova
password = NOVA_PASS

[cinder]
...
```

```
memcached_servers = controller：11211
region_name = RegionOne
auth_uri = http：//controller：5000
auth_url = http：//controller：35357
auth_type = password
project_domain_name = default
project_name = service
user_domain_name = default
username = cinder
password = CINDER_PASS
```

c. 在［generic］部分，配置 generic 驱动：

由于初始配置文件中并没有［generic］部分，因此可以在文件尾部添加此部分内容。

```
[generic]
share_backend_name = GENERIC
share_driver = manila.share.drivers.generic.GenericShareDriver
driver_handles_share_servers = True
service_instance_flavor_id = 100
service_image_name = manila-service-image
service_instance_user = manila
service_instance_password = manila
interface_driver = manila.network.linux.interface.BridgeInterfaceDriver
```

11.4.4　完成安装

启动文件共享存储服务，并将其配置为随系统启动：

systemctl enable openstack-manila-share.service
systemctl start openstack-manila-share.service

11.5　验证操作（控制节点）

1) 获得 admin 凭证来获取只有管理员才能执行的命令的访问权限

./root/admin-openrc

2) 列出服务组件，以验证是否每个进程都成功启动

manila service-list

模式 1：没有驱动支持的文件共享服务器管理，如图 11-14 所示。

```
[root@controller ~]# manila service-list
+----+------------------+---------------------+------+---------+-------+----------------------------+
| Id | Binary           | Host                | Zone | Status  | State | Updated_at                 |
+----+------------------+---------------------+------+---------+-------+----------------------------+
| 1  | manila-scheduler | controller          | nova | enabled | up    | 2017-03-27T08:45:44.000000 |
| 2  | manila-share     | blockstorage01@lvm  | nova | enabled | up    | 2017-03-27T08:45:09.000000 |
+----+------------------+---------------------+------+---------+-------+----------------------------+
```

图 11-14　无驱动支持的文件共享服务器管理

在正确配置的情况下，输出结果应该包括控制节点上的一个代理和每个文件共享存储节点上的一个代理，文件共享存储节点名显示为 HOSTNAME@lvm。

模式 2：有驱动支持的文件共享服务器管理，如图 11-15 所示。

```
[root@controller ~]# manila service-list
+----+------------------+-------------------------+------+---------+-------+----------------------------+
| Id | Binary           | Host                    | Zone | Status  | State | Updated_at                 |
+----+------------------+-------------------------+------+---------+-------+----------------------------+
| 1  | manila-scheduler | controller              | nova | enabled | up    | 2017-03-27T08:57:43.000000 |
| 2  | manila-share     | blockstorage01@generic  | nova | enabled | up    | 2017-03-27T08:57:04.000000 |
+----+------------------+-------------------------+------+---------+-------+----------------------------+
```

图 11-15　有驱动支持的文件共享服务器管理

在正确配置的情况下，输出结果应该包括控制节点上的一个代理和每个文件共享存储节点上的一个代理，文件共享存储节点名显示为 HOSTNAME@generic。

第 12 章 对象存储（Swift）服务

12.1 服务概述

对象存储是一个多租户的对象存储系统，它支持大规模扩展，可以通过 RESTful HTTP 应用程序接口实现大型的非结构化数据的低成本管理。

对象存储服务（swift）构建在标准硬件存储基础设施之上，无须采用 RAID（磁盘冗余阵列），可以通过 REST API 提供对象存储及恢复。该服务在软件层面引入一致性散列技术，用以提高数据的冗余性、高可用性和可伸缩性，支持多租户模式、容器和对象读写操作，适合解决互联网应用场景下的非结构化数据存储问题。

Swift 不能像传统文件系统那样进行挂载和访问，只能通过 Rest API 接口访问数据，并且这些 API 与亚马逊的 S3 服务 API 是兼容的。Swift 不同于传统文件系统和实时数据存储系统，它适用于存储、获取一些静态的、永久性的数据，并在需要的时候进行更新。在 OpenStack 中，Swift 主要应用于存储虚拟机镜像，用于 Glance 的后端存储。在实际运用中，Swift 常被运用于网盘系统，多为存储图片、邮件、视频等静态资源。

本章讲述如何安装和配置在存储节点上处理 account、container 和 object 服务请求的代理服务。为简单起见，本书只介绍在控制节点上的安装和配置代理服务。不过，也可以在任何与存储节点网络联通的节点上运行代理服务，另外还可以在多个节点上安装和配置代理服务提高性能和冗余。

对象存储服务包含下列组件：

1. 代理服务器（swift-proxy-server）

接收 OpenStack 对象存储 API 和纯粹的 HTTP 请求以上传文件、更改元数据以及创建容器。它可在 Web 浏览器下显示文件和容器列表。为了改进性能，代理服务可以使用可选的缓存，通常部署的是 memcache。

2. 账户服务器（swift-account-server）

管理由对象存储定义的账户。

3. 容器服务器（swift-container-server）

管理容器或文件夹的映射，对象存储内部。

4. 对象服务器（swift-object-server）

在存储节点上管理实际的对象，比如：文件。

5. 各种定期进程

为了驾驭大型数据存储的任务,复制服务需要在集群内确保一致性和可用性,其他定期进程有审计、更新和 reaper。

6. WSGI 中间件

掌控认证,使用 OpenStack 认证服务。

7. swift 客户端

用户可以通过此命令行客户端来向 REST API 提交命令,授权的用户角色可以是管理员用户、经销商用户、或者是 swift 用户。

8. swift-init

初始化环链文件生成的脚本,将守护进程名称当作参数并提供命令。

9. swift-recon

一个用于检索集群不同度量与计量信息的命令行工具,已经被 swift-recon 中间件采用。

10. swift-ring-builder

存储环链建立并重平衡的实用程序。

12.2 安装前准备(控制节点)

代理服务依赖于诸如身份认证服务所提供的认证和授权机制。但是与其他服务不同,它也提供了一个内部机制可以在没有任何其他 OpenStack 服务的情况下运行。不过为了简单起见,本书引用 keystone 中的身份认证服务。在配置对象存储服务前,必须创建服务凭证和 API 端点。

对象存储服务不是使用控制节点上的 SQL 数据库,而是使用在每个存储节点上的分布式 SQL 数据库。

12.2.1 创建用户、服务实体和 API 端点

1. 获得 admin 凭证

获得 admin 凭证来获取只有管理员才能执行的命令的访问权限:

. /root/admin-openrc

2. 创建 swift 用户

1)创建 swift 用户

openstack user create--domain default \
    --password SWIFT_PASS swift

本书示例中将 swift 用户密码设置为 SWIFT_PASS,如图 12-1 所示。

```
[root@controller ~]# openstack user create --domain default \
> --password SWIFT_PASS swift
+-----------+----------------------------------+
| Field     | Value                            |
+-----------+----------------------------------+
| domain_id | 1224d01d12d5451191c2f0aeefa6ad6a |
| enabled   | True                             |
| id        | e48405c26aaf425c990d9fcf5696cb48 |
| name      | swift                            |
+-----------+----------------------------------+
```

图 12 - 1　创建 swift 用户

2）添加 admin 角色到 swift 用户和 service 项目上

openstack role add--project service--user swift admin

如图 12 - 2 所示。

```
[root@controller ~]# openstack role add --project service --user swift admin
[root@controller ~]#
```

图 12 - 2　添加 admin 角色到 swift 用户和 service 项目上

3. 创建服务实体和身份认证服务

创建 swift 服务实体：

openstack service create--name swift \
　--description " OpenStack Object Storage" object-store

如图 12 - 3 所示。

```
[root@controller ~]# openstack service create --name swift \
> --description "OpenStack Object Storage" object-store
+-------------+----------------------------------+
| Field       | Value                            |
+-------------+----------------------------------+
| description | OpenStack Object Storage         |
| enabled     | True                             |
| id          | 631d9ab7e4d3496f95948217236a4cbb |
| name        | swift                            |
| type        | object-store                     |
+-------------+----------------------------------+
```

图 12 - 3　创建 swift 服务实体

4. 创建 API 端点

创建文件共享服务 API 端点：

对象存储服务每个服务实体都需要端点，分别如图 12 - 4 到图 12 - 6 所示。

openstack endpoint create--region RegionOne \
　object-store public http：//controller：8080/v1/AUTH_ % \（tenant_id \）s

```
[root@controller ~]# openstack endpoint create --region RegionOne \
> object-store public http://controller:8080/v1/AUTH_%\(tenant_id\)s
+--------------+-------------------------------------------+
| Field        | Value                                     |
+--------------+-------------------------------------------+
| enabled      | True                                      |
| id           | e809d61ef62843bf806f78c27d27fdd7          |
| interface    | public                                    |
| region       | RegionOne                                 |
| region_id    | RegionOne                                 |
| service_id   | 631d9ab7e4d3496f95948217236a4cbb          |
| service_name | swift                                     |
| service_type | object-store                              |
| url          | http://controller:8080/v1/AUTH_%(tenant_id)s |
+--------------+-------------------------------------------+
```

图 12-4 创建 API 端点 (public)

openstack endpoint create--region RegionOne \

object-store internal http：//controller：8080/v1/AUTH_% \ （tenant_id \ ) s

```
[root@controller ~]# openstack endpoint create --region RegionOne \
> object-store internal http://controller:8080/v1/AUTH_%\(tenant_id\)s
+--------------+-------------------------------------------+
| Field        | Value                                     |
+--------------+-------------------------------------------+
| enabled      | True                                      |
| id           | b87fa5076f284b75996e11119ae69d7d          |
| interface    | internal                                  |
| region       | RegionOne                                 |
| region_id    | RegionOne                                 |
| service_id   | 631d9ab7e4d3496f95948217236a4cbb          |
| service_name | swift                                     |
| service_type | object-store                              |
| url          | http://controller:8080/v1/AUTH_%(tenant_id)s |
+--------------+-------------------------------------------+
```

图 12-5 创建 API 端点 (internal)

openstack endpoint create--region RegionOne \

object-store admin http：//controller：8080/v1

```
[root@controller ~]# openstack endpoint create --region RegionOne \
> object-store admin http://controller:8080/v1
+--------------+----------------------------------+
| Field        | Value                            |
+--------------+----------------------------------+
| enabled      | True                             |
| id           | ccfe52d9a68947a6a2f0d4a615ac4916 |
| interface    | admin                            |
| region       | RegionOne                        |
| region_id    | RegionOne                        |
| service_id   | 631d9ab7e4d3496f95948217236a4cbb |
| service_name | swift                            |
| service_type | object-store                     |
| url          | http://controller:8080/v1        |
+--------------+----------------------------------+
```

图 12-6 创建 API 端点 (admin)

12.3 安装和配置（控制节点）

12.3.1 安装软件包

1）安装 swift 软件包

yum -y install openstack-swift-proxy python-swiftclient \
　　python-keystoneclient python-keystonemiddleware \
　　memcached

2）获取代理服务的配置文件

当主机可访问互联网时，可以从对象存储的在线仓库源中下载配置文件。

curl -o /etc/swift/proxy-server.conf \
　　https://git.openstack.org/cgit/openstack/swift/plain/etc/proxy-server.conf-sample?h=stable/mitaka

在本书示例中将使用预先提供的配置文件，并将其复制到目标位置。

\ cp /openstack/data/proxy-server.conf-sample /etc/swift/proxy-server.conf

13.3.2 配置代理服务

编辑 /etc/swift/proxy-server.conf 文件：

由于默认配置文件在各发行版本中可能不同，因此，在进行修改的同时可能需要添加部分选项。另外，在配置片段中的省略号（...）表示默认的配置选项，应该保留。

gedit /etc/swift/proxy-server.conf

a. 在［DEFAULT］部分，配置绑定端口，用户和配置目录：

[DEFAULT]
...
bind_port = 8080
swift_dir = /etc/swift
user = swift

b. 在［pipeline：main］部分，删除 tempurl 和 tempauth 模块并增加 authtoken 和 keystoneauth 模块，不要改变模块的顺序：

[pipeline：main]
pipeline = catch_errors gatekeeper healthcheck proxy-logging cache container_sync bulk ratelimit authtoken keystoneauth container-quotas account-quotas slo dlo versioned_writes proxy-logging proxy-server

c. 在 [app：proxy-server] 部分，启动自动账户创建：

[app：proxy-server]
use = egg：swift#proxy
...
account_autocreate = True

d. 在 [filter：keystoneauth] 部分，配置操作员角色：
默认配置文件中 [filter：keystoneauth] 项会被注释掉，注意取消注释。

[filter：keystoneauth]
use = egg：swift#keystoneauth
...
operator_roles = admin，user

e. 在 [filter：authtoken] 部分，配置认证服务访问：
默认配置文件中 [filter：authtoken] 项会被注释掉，注意取消注释。

[filter：authtoken]
paste.filter_factory = keystonemiddleware.auth_token：filter_factory
...
auth_uri = http：//controller：5000
auth_url = http：//controller：35357
memcached_servers = controller：11211
auth_type = password
project_domain_name = default
user_domain_name = default
project_name = service
username = swift
password = SWIFT_PASS
delay_auth_decision = True

注释或者删除掉在 [filter：authtoken] 部分的所有其他的内容。
f. 在 [filter：cache] 部分，配置 memcached 的位置：

```
[filter: cache]
use = egg: swift#memcache
...
memcache_servers = controller: 11211
```

12.4 安装和配置（对象存储节点）

本节为操作账号、容器和对象服务安装和配置存储节点。为简单起见，这里配置两个存储节点，每个包含两个空本地块存储设备，使用的是 /dev/sdb 和 /dev/sdc，可以用不同的值代替的特定节点。

尽管对象存储通过 extended attributes（xattr）支持所有文件系统，但测试表明使用 XFS 时性能最好，可靠性最高。

在每个存储节点上执行这些步骤。

12.4.1 安装前准备

1）安装支持工具软件包

yum -y install xfsprogs rsync

2）使用 XFS 格式化 /dev/sdb 和 /dev/sdc 设备

mkfs.xfs /dev/sdb

mkfs.xfs /dev/sdc

如图 12-7 所示。

```
[root@objectstorage01 ~]# mkfs.xfs /dev/sdb
meta-data=/dev/sdb              isize=512    agcount=4, agsize=655360 blks
         =                       sectsz=512   attr=2, projid32bit=1
         =                       crc=1        finobt=0, sparse=0
data     =                       bsize=4096   blocks=2621440, imaxpct=25
         =                       sunit=0      swidth=0 blks
naming   =version 2              bsize=4096   ascii-ci=0 ftype=1
log      =internal log           bsize=4096   blocks=2560, version=2
         =                       sectsz=512   sunit=0 blks, lazy-count=1
realtime =none                   extsz=4096   blocks=0, rtextents=0
[root@objectstorage01 ~]#
[root@objectstorage01 ~]# mkfs.xfs /dev/sdc
meta-data=/dev/sdc              isize=512    agcount=4, agsize=655360 blks
         =                       sectsz=512   attr=2, projid32bit=1
         =                       crc=1        finobt=0, sparse=0
data     =                       bsize=4096   blocks=2621440, imaxpct=25
         =                       sunit=0      swidth=0 blks
naming   =version 2              bsize=4096   ascii-ci=0 ftype=1
log      =internal log           bsize=4096   blocks=2560, version=2
         =                       sectsz=512   sunit=0 blks, lazy-count=1
realtime =none                   extsz=4096   blocks=0, rtextents=0
[root@objectstorage01 ~]#
```

图 12-7 使用 XFS 格式化 /dev/sdb 和 /dev/sdc 设备

3）创建挂载点目录结构

\# mkdir -p /srv/node/sdb

\# mkdir -p /srv/node/sdc

4）编辑 /etc/fstab 文件

\# gedit /etc/fstab

在文件尾添加以下内容：

/dev/sdb /srv/node/sdb xfs noatime, nodiratime, nobarrier, logbufs = 8 0 2
/dev/sdc /srv/node/sdc xfs noatime, nodiratime, nobarrier, logbufs = 8 0 2

5）挂载设备

\# mount /srv/node/sdb

\# mount /srv/node/sdc

6）创建 /etc/rsyncd.conf 文件

\# gedit /etc/rsyncd.conf

在文件尾添加以下内容。

uid = swift
gid = swift
log file = /var/log/rsyncd.log
pid file = /var/run/rsyncd.pid
address = MANAGEMENT_INTERFACE_IP_ADDRESS

[account]
max connections = 2
path = /srv/node/
read only = False
lock file = /var/lock/account.lock

[container]
max connections = 2
path = /srv/node/
read only = False
lock file = /var/lock/container.lock

[object]

```
max connections = 2
path = /srv/node/
read only = False
lock file = /var/lock/object.lock
```

由于 rsync 服务不需要认证,因此考虑将它安装在私有网络的环境中。

将 MANAGEMENT_INTERFACE_IP_ADDRESS 替换为对象存储节点的管理网络 IP 地址,本书示例中将第 1 个对象存储节点设置为 192.168.1.51,将第 2 个对象存储节点设置为 192.168.1.61。

12.4.2 安装和配置

请在每个对象存储节点上执行本节中的步骤。

12.4.3 安装软件包

1)安装 swift 软件包

```
# yum -y install openstack-swift-account openstack-swift-container \
    openstack-swift-object
```

2)获取 accounting、container 以及 object 服务的配置文件

当主机可访问互联网时,可以从对象存储的在线仓库源中下载配置文件。

```
# curl -o /etc/swift/account-server.conf \
    https://git.openstack.org/cgit/openstack/swift/plain/etc/account-server.conf-sample?h=stable/mitaka
# curl -o /etc/swift/container-server.conf \
    https://git.openstack.org/cgit/openstack/swift/plain/etc/container-server.conf-sample?h=stable/mitaka
# curl -o /etc/swift/object-server.conf \
    https://git.openstack.org/cgit/openstack/swift/plain/etc/object-server.conf-sample?h=stable/mitaka
```

在本书示例中将使用预先提供的配置文件,并将其复制到目标位置。

```
# \cp /openstack/data/account-server.conf-sample \
    /etc/swift/account-server.conf
# \cp /openstack/data/container-server.conf-sample \
    /etc/swift/container-server.conf
# \cp /openstack/data/object-server.conf-sample \
    /etc/swift/object-server.conf
```

12.4.4 配置账户服务

编辑 /etc/swift/account-server.conf 文件：

由于默认配置文件在各发行版本中可能不同，因此，在进行修改的同时可能需要添加部分选项。另外，在配置片段中的省略号（...）表示默认的配置选项，应该保留。

gedit /etc/swift/account-server.conf

a. 在［DEFAULT］部分，配置绑定 IP 地址、绑定端口、用户、配置目录和挂载目录：

```
[DEFAULT]
...
bind_ip = MANAGEMENT_INTERFACE_IP_ADDRESS
bind_port = 6002
user = swift
swift_dir = /etc/swift
devices = /srv/node
mount_check = True
```

将 MANAGEMENT_INTERFACE_IP_ADDRESS 替换为对象存储节点的管理网络 IP 地址，本书示例中将第 1 个对象存储节点设置为 192.168.1.51，将第 2 个对象存储节点设置为 192.168.1.61。

b. 在［pipeline：main］部分，启用合适的模块：

```
[pipeline:main]
...
pipeline = healthcheck recon account-server
```

c. 在［filter：recon］部分，配置 recon（meters）缓存目录：

```
[filter:recon]
use = egg:swift#recon
...
recon_cache_path = /var/cache/swift
```

12.4.5 配置容器服务

编辑 /etc/swift/container-server.conf 文件：

由于默认配置文件在各发行版本中可能不同，因此，在进行修改的同时可能需要添加部分选项。另外，在配置片段中的省略号（…）表示默认的配置选项，应该保留。

gedit /etc/swift/container-server.conf

a. 在［DEFAULT］部分，配置绑定 IP 地址、绑定端口、用户、配置目录和挂载目录：

```
[DEFAULT]
...
bind_ip = MANAGEMENT_INTERFACE_IP_ADDRESS
bind_port = 6001
user = swift
swift_dir = /etc/swift
devices = /srv/node
mount_check = True
```

将 MANAGEMENT_INTERFACE_IP_ADDRESS 替换为对象存储节点的管理网络 IP 地址，在本书示例中将第 1 个对象存储节点设置为 192.168.1.51，将第 2 个对象存储节点设置为 192.168.1.61。

b. 在［pipeline：main］部分，启用合适的模块：

```
[pipeline:main]
...
pipeline = healthcheck recon container-server
```

c. 在［filter：recon］部分，配置 recon（meters）缓存目录：

```
[filter:recon]
use = egg:swift#recon
...
recon_cache_path = /var/cache/swift
```

12.4.6 配置对象服务

编辑 /etc/swift/object-server.conf 文件：

由于默认配置文件在各发行版本中可能不同。因此，在进行修改的同时可能需要添加部分选项。另外，在配置片段中的省略号（…）表示默认的配置选项，应该保留。

gedit /etc/swift/object-server.conf

a. 在［DEFAULT］部分，配置绑定 IP 地址、绑定端口、用户、配置目录和挂载目录：

[DEFAULT]
...
bind_ip = MANAGEMENT_INTERFACE_IP_ADDRESS
bind_port = 6000
user = swift
swift_dir = /etc/swift
devices = /srv/node
mount_check = True

将 MANAGEMENT_INTERFACE_IP_ADDRESS 替换为对象存储节点的管理网络 IP 地址，本书示例中将第 1 个对象存储节点设置为 192.168.1.51，将第 2 个对象存储节点设置为 192.168.1.61。

b. 在［pipeline：main］部分，启用合适的模块：

[pipeline：main]
...
pipeline = healthcheck recon object-server

c. 在［filter：recon］部分，配置 recon（meters）缓存目录和 lock 目录：

[filter：recon]
use = egg：swift#recon
...
recon_cache_path = /var/cache/swift
recon_lock_path = /var/lock

12.4.7 配置目录权限

1）为挂载目录赋予特定的权限

chown-R swift：swift /srv/node

2）创建 recon（meters）缓存目录，并赋予特定的权限

mkdir-p /var/cache/swift
chown-R root：swift /var/cache/swift

chmod-R 775 /var/cache/swift

12.5 创建、分发并初始化 rings（控制节点）

在开始启动对象存储服务前，必须创建初始化 account、container 和 object rings。ring builder 创建每个节点用户决定和部署存储体系的配置文件。

简单地说，本书示例中使用一个 region 和包括两个最多 2^10（1024）个分区的 zone、每个对象的 3 个副本和移动分区超过 1 次时最少 1 个小时时长。对于对象存储而言，一个分区意味着存储设备的一个目录而不是传统的分区表。

12.5.1 创建账户 ring

账户服务器使用账户 ring 来维护一个容器的列表。

1）切换到 /etc/swift 目录

cd /etc/swift

2）创建基本 account.builder 文件

swift-ring-builder account.builder create 10 3 1

如图 12-8 所示。

```
[root@controller swift]# swift-ring-builder account.builder create 10 3 1
[root@controller swift]#
```

图 12-8 创建基本 account.builder 文件

3）添加每个节点到 ring 中

为每个存储节点上面重复执行这个命令。在这个例子的架构中，使用该命令的四个变量，/dev/sdb 存储设备，大小为 10：

swift-ring-builder account.builder add \
  --region 1--zone 1--ip 192.168.1.51--port 6002--device sdb--weight 10
swift-ring-builder account.builder add \
  --region 1--zone 1--ip 192.168.1.51--port 6002--device sdc--weight 10
swift-ring-builder account.builder add \
  --region 1--zone 2--ip 192.168.1.61--port 6002--device sdb--weight 10
swift-ring-builder account.builder add \
  --region 1--zone 2--ip 192.168.1.61--port 6002--device sdc--weight 10

如图 12-9 所示。

```
[root@controller swift]#
[root@controller swift]# swift-ring-builder account.builder add \
> --region 1 --zone 1 --ip 192.168.1.51 --port 6002 --device sdb --weight 10
Device d0r1z1-192.168.1.51:6002R192.168.1.51:6002/sdb_"" with 10.0 weight got id 0
[root@controller swift]#
[root@controller swift]# swift-ring-builder account.builder add \
> --region 1 --zone 1 --ip 192.168.1.51 --port 6002 --device sdc --weight 10
Device d1r1z1-192.168.1.51:6002R192.168.1.51:6002/sdc_"" with 10.0 weight got id 1
[root@controller swift]#
[root@controller swift]# swift-ring-builder account.builder add \
> --region 1 --zone 2 --ip 192.168.1.61 --port 6002 --device sdb --weight 10
Device d2r1z2-192.168.1.61:6002R192.168.1.61:6002/sdb_"" with 10.0 weight got id 2
[root@controller swift]#
[root@controller swift]# swift-ring-builder account.builder add \
> --region 1 --zone 2 --ip 192.168.1.61 --port 6002 --device sdc --weight 10
Device d3r1z2-192.168.1.61:6002R192.168.1.61:6002/sdc_"" with 10.0 weight got id 3
```

图 12-9 添加每个节点到 ring 中

4）验证 ring 的内容

swift-ring-builder account.builder

如图 12-10 所示。

```
[root@controller swift]# swift-ring-builder account.builder
account.builder, build version 4
1024 partitions, 3.000000 replicas, 1 regions, 2 zones, 4 devices, 100.00 balance, 0.00 dispersion
The minimum number of hours before a partition can be reassigned is 1 (0:00:00 remaining)
The overload factor is 0.00% (0.000000)
Ring file account.ring.gz not found, probably it hasn't been written yet
Devices:    id  region  zone      ip address  port  replication ip  replication port      name weight partitions
 balance flags meta
             0       1     1    192.168.1.51  6002    192.168.1.51              6002       sdb  10.00          0
-100.00
             1       1     1    192.168.1.51  6002    192.168.1.51              6002       sdc  10.00          0
-100.00
             2       1     2    192.168.1.61  6002    192.168.1.61              6002       sdb  10.00          0
-100.00
             3       1     2    192.168.1.61  6002    192.168.1.61              6002       sdc  10.00          0
-100.00
```

图 12-10 验证 ring 的内容

5）平衡 ring

swift-ring-builder account.builder rebalance

如图 12-11 所示。

```
[root@controller swift]# swift-ring-builder account.builder rebalance
Reassigned 3072 (300.00%) partitions. Balance is now 0.00.  Dispersion is now 0.00
```

图 12-11 平衡 ring

12.5.2 创建容器 ring

容器服务器使用容器环来维护对象的列表。但是它不跟踪对象的位置。

1）切换到 /etc/swift 目录

cd /etc/swift

2）创建基本 container. builder 文件

swift-ring-builder container. builder create 10 3 1

如图 12-12 所示。

```
[root@controller swift]# swift-ring-builder container.builder create 10 3 1
[root@controller swift]#
```

图 12-12　创建基本 container. builder 文件

3）添加每个节点到 ring 中

在每个存储节点上面重复执行这个命令。在这个例子的架构中，使用该命令的四个变量：

swift-ring-builder container. builder add \
　--region 1--zone 1--ip 192. 168. 1. 51--port 6001--device sdb--weight 10
swift-ring-builder container. builder add \
　--region 1--zone 1--ip 192. 168. 1. 51--port 6001--device sdc--weight 10
swift-ring-builder container. builder add \
　--region 1--zone 2--ip 192. 168. 1. 61--port 6001--device sdb--weight 10
swift-ring-builder container. builder add \
　--region 1--zone 2--ip 192. 168. 1. 61--port 6001--device sdc--weight 10

如图 12-13 所示。

```
[root@controller swift]# swift-ring-builder container.builder add \
> --region 1 --zone 1 --ip 192.168.1.51 --port 6001 --device sdb --weight 10
Device d0r1z1-192.168.1.51:6001R192.168.1.51:6001/sdb_"" with 10.0 weight got id 0
[root@controller swift]#
[root@controller swift]# swift-ring-builder container.builder add \
> --region 1 --zone 1 --ip 192.168.1.51 --port 6001 --device sdc --weight 10
Device d1r1z1-192.168.1.51:6001R192.168.1.51:6001/sdc_"" with 10.0 weight got id 1
[root@controller swift]#
[root@controller swift]# swift-ring-builder container.builder add \
> --region 1 --zone 2 --ip 192.168.1.61 --port 6001 --device sdb --weight 10
Device d2r1z2-192.168.1.61:6001R192.168.1.61:6001/sdb_"" with 10.0 weight got id 2
[root@controller swift]#
[root@controller swift]# swift-ring-builder container.builder add \
> --region 1 --zone 2 --ip 192.168.1.61 --port 6001 --device sdc --weight 10
Device d3r1z2-192.168.1.61:6001R192.168.1.61:6001/sdc_"" with 10.0 weight got id 3
```

图 12-13　添加每个节点到 ring 中

4）验证 ring 的内容

swift-ring-builder container. builder

如图 12-14 所示。

```
[root@controller swift]# swift-ring-builder container.builder
container.builder, build version 4
1024 partitions, 3.000000 replicas, 1 regions, 2 zones, 4 devices, 100.00 balance, 0.00 dispersion
The minimum number of hours before a partition can be reassigned is 1 (0:00:00 remaining)
The overload factor is 0.00% (0.000000)
Ring file container.ring.gz not found, probably it hasn't been written yet
Devices:    id  region  zone      ip address  port  replication ip  replication port      name weight partitions
 balance flags meta
             0       1     1    192.168.1.51  6001    192.168.1.51            6001        sdb  10.00         0
-100.00
             1       1     1    192.168.1.51  6001    192.168.1.51            6001        sdc  10.00         0
-100.00
             2       1     2    192.168.1.61  6001    192.168.1.61            6001        sdb  10.00         0
-100.00
             3       1     2    192.168.1.61  6001    192.168.1.61            6001        sdc  10.00         0
-100.00
```

图 12 – 14　验证 ring 的内容

5）平衡 ring

\# swift-ring-builder container.builder rebalance

如图 12 – 15 所示。

```
[root@controller swift]# swift-ring-builder container.builder rebalance
Reassigned 3072 (300.00%) partitions. Balance is now 0.00.  Dispersion is now 0.00
```

图 12 – 15　平衡 ring

12.5.3　创建对象 ring

对象服务器使用对象环来维护对象在本地设备上的位置列表。

1）切换到 /etc/swift 目录

\# cd /etc/swift

2）创建基本 object.builder 文件

\# swift-ring-builder object.builder create 10 3 1

如图 12 – 16 所示。

```
[root@controller swift]# swift-ring-builder object.builder create 10 3 1
[root@controller swift]#
```

图 12 – 16　创建基本 object.builder 文件

3）添加每个节点到 ring 中：

\# swift-ring-builder object.builder add \
　--region 1--zone 1--ip 192.168.1.51--port 6000--device sdb--weight 10

\# swift-ring-builder object.builder add \
　--region 1--zone 1--ip 192.168.1.51--port 6000--device sdc--weight 10

\# swift-ring-builder object.builder add \
　--region 1--zone 2--ip 192.168.1.61--port 6000--device sdb--weight 10

swift-ring-builder object.builder add \
  --region 1--zone 2--ip 192.168.1.61--port 6000--device sdc--weight 10

如图 12-17 所示。

```
[root@controller swift]#
[root@controller swift]# swift-ring-builder object.builder add \
> --region 1 --zone 1 --ip 192.168.1.51 --port 6000 --device sdb --weight 10
Device d0r1z1-192.168.1.51:6000R192.168.1.51:6000/sdb_"" with 10.0 weight got id 0
[root@controller swift]#
[root@controller swift]# swift-ring-builder object.builder add \
> --region 1 --zone 1 --ip 192.168.1.51 --port 6000 --device sdc --weight 10
Device d1r1z1-192.168.1.51:6000R192.168.1.51:6000/sdc_"" with 10.0 weight got id 1
[root@controller swift]#
[root@controller swift]# swift-ring-builder object.builder add \
> --region 1 --zone 2 --ip 192.168.1.61 --port 6000 --device sdb --weight 10
Device d2r1z2-192.168.1.61:6000R192.168.1.61:6000/sdb_"" with 10.0 weight got id 2
[root@controller swift]#
[root@controller swift]# swift-ring-builder object.builder add \
> --region 1 --zone 2 --ip 192.168.1.61 --port 6000 --device sdc --weight 10
Device d3r1z2-192.168.1.61:6000R192.168.1.61:6000/sdc_"" with 10.0 weight got id 3
```

图 12-17 添加每个节点到 ring

4）验证 ring 的内容

swift-ring-builder object.builder

如图 12-18 所示。

```
[root@controller swift]# swift-ring-builder object.builder
object.builder, build version 4
1024 partitions, 3.000000 replicas, 1 regions, 2 zones, 4 devices, 100.00 balance, 0.00 dispersion
The minimum number of hours before a partition can be reassigned is 1 (0:00:00 remaining)
The overload factor is 0.00% (0.000000)
Ring file object.ring.gz not found, probably it hasn't been written yet
Devices:    id  region  zone      ip address  port  replication ip  replication port  name weight partitions
    balance flags meta
             0       1     1    192.168.1.51  6000    192.168.1.51              6000  sdb   10.00          0
    -100.00
             1       1     1    192.168.1.51  6000    192.168.1.51              6000  sdc   10.00          0
    -100.00
             2       1     2    192.168.1.61  6000    192.168.1.61              6000  sdb   10.00          0
    -100.00
             3       1     2    192.168.1.61  6000    192.168.1.61              6000  sdc   10.00          0
    -100.00
```

图 12-18 验证 ring 的内容

5）平衡 ring

swift-ring-builder object.builder rebalance

如图 12-19 所示。

```
[root@controller swift]# swift-ring-builder object.builder rebalance
Reassigned 3072 (300.00%) partitions. Balance is now 0.00.  Dispersion is now 0.00
```

图 12-19 平衡 ring

12.5.4 分发环配置文件

将控制节点 /etc/swift 目录中的 account.ring.gz、container.ring.gz 和 object.ring.gz 文件复制到每个存储节点和其他运行了代理服务的额外节点的 /etc/swift 目录。

scp /etc/swift/*.ring.gz root@192.168.1.51：/etc/swift/

scp /etc/swift/*.ring.gz root@192.168.1.61：/etc/swift/

当确认继续时，请先输入"yes"，然后根据提示输入对象存储节点的 root 用户密码，如图 12-20 所示。

```
[root@controller swift]# scp /etc/swift/*.ring.gz root@192.168.1.51:/etc/swift/
The authenticity of host '192.168.1.51 (192.168.1.51)' can't be established.
ECDSA key fingerprint is 71:46:58:6b:01:12:e7:fb:28:ac:93:a9:c6:1b:2e:33.
Are you sure you want to continue connecting (yes/no)? yes
Warning: Permanently added '192.168.1.51' (ECDSA) to the list of known hosts.
root@192.168.1.51's password:
account.ring.gz                              100% 1462     1.4KB/s   00:00
container.ring.gz                            100% 1466     1.4KB/s   00:00
object.ring.gz                               100% 1470     1.4KB/s   00:00
[root@controller swift]#
[root@controller swift]# scp /etc/swift/*.ring.gz root@192.168.1.61:/etc/swift/
The authenticity of host '192.168.1.61 (192.168.1.61)' can't be established.
ECDSA key fingerprint is 71:46:58:6b:01:12:e7:fb:28:ac:93:a9:c6:1b:2e:33.
Are you sure you want to continue connecting (yes/no)? yes
Warning: Permanently added '192.168.1.61' (ECDSA) to the list of known hosts.
root@192.168.1.61's password:
account.ring.gz                              100% 1462     1.4KB/s   00:00
container.ring.gz                            100% 1466     1.4KB/s   00:00
object.ring.gz                               100% 1470     1.4KB/s   00:00
```

图 12-20　分发环配置文件

12.6　完成安装（控制节点、对象存储节点）

1) 在控制节点上，获取 swift 服务的配置文件

当主机可访问互联网时，可以从对象存储的在线仓库源中下载配置文件。

curl-o /etc/swift/swift.conf \

　　https：//git.openstack.org/cgit/openstack/swift/plain/etc/swift.conf-sample?h=stable/mitaka

在本书示例中将使用预先提供的配置文件，并将其复制到目标位置。

\ cp /openstack/data/swift.conf-sample /etc/swift/swift.conf

2) 在控制节点上，编辑 /etc/swift/swift.conf 文件

由于默认配置文件在各发行版本中可能不同，因此，在进行修改的同时可能需要添加部分选项。另外，在配置片段中的省略号（...）表示默认的配置选项，应该保留。

gedit /etc/swift/swift.conf

a. 在 [swift-hash] 部分，为你的环境配置哈希路径前缀和后缀：

[swift-hash]

...

swift_hash_path_suffix = HASH_PATH_SUFFIX

swift_hash_path_prefix = HASH_PATH_PREFIX

建议使用 openssl rand-hex 10 | md5sum 命令生成随机且复杂的哈希值，来替换 HASH_PATH_SUFFIX 和 HASH_PATH_PREFIX。

为简单起见，本书示例中将不替换 HASH_PATH_SUFFIX 和 HASH_PATH_PREFIX。

b. 在［storage-policy：0］部分，配置默认存储策略：

[storage-policy：0]

...

name = Policy-0

default = yes

3）在控制节点上，复制 swift.conf 文件到每个对象存储节点和其他允许了代理服务的额外节点的 /etc/swift 目录

scp /etc/swift/swift.conf root@192.168.1.51：/etc/swift/

scp /etc/swift/swift.conf root@192.168.1.61：/etc/swift/

当远程连接到对象存储节点时，根据提示输入对象存储节点的 root 用户密码，如图 12-21 所示。

```
[root@controller swift]# scp /etc/swift/swift.conf root@192.168.1.51:/etc/swift/
root@192.168.1.51's password:
swift.conf                                                100%  7569     7.4KB/s   00:00
[root@controller swift]#
[root@controller swift]# scp /etc/swift/swift.conf root@192.168.1.61:/etc/swift/
root@192.168.1.61's password:
swift.conf                                                100%  7569     7.4KB/s   00:00
```

图 12-21 输入 root 密码

4）在所有节点（**控制节点和对象存储节点**）上，确认配置文件目录是否有合适的所有权

chown-R root：swift /etc/swift

5）在控制节点和其他运行了代理服务的节点上，启动对象存储代理服务及其依赖服务，并将它们配置为随系统启动

systemctl enable openstack-swift-proxy.service memcached.service

systemctl start openstack-swift-proxy.service memcached.service

6）在对象存储节点上，启动对象存储服务的账户服务，并将其配置为随系统启动

systemctl enable openstack-swift-account.service \

```
    openstack-swift-account-auditor.service \
    openstack-swift-account-reaper.service \
    openstack-swift-account-replicator.service
# systemctl start openstack-swift-account.service \
    openstack-swift-account-auditor.service \
    openstack-swift-account-reaper.service \
    openstack-swift-account-replicator.service
```

7）在对象存储节点上，启动对象存储服务的容器服务，并将其配置为随系统启动

```
# systemctl enable openstack-swift-container.service \
    openstack-swift-container-auditor.service \
    openstack-swift-container-replicator.service \
    openstack-swift-container-updater.service
# systemctl start openstack-swift-container.service \
    openstack-swift-container-auditor.service \
    openstack-swift-container-replicator.service \
    openstack-swift-container-updater.service
```

8）在对象存储节点上，启动对象存储服务的对象服务，并将其配置为随系统启动

```
# systemctl enable openstack-swift-object.service \
    openstack-swift-object-auditor.service \
    openstack-swift-object-replicator.service \
    openstack-swift-object-updater.service
# systemctl start openstack-swift-object.service \
    openstack-swift-object-auditor.service \
    openstack-swift-object-replicator.service \
    openstack-swift-object-updater.service
```

12.7 验证操作（控制节点）

1）获得admin凭证来获取只有管理员才能执行的命令的访问权限

```
# ./root/admin-openrc
```

2）列出对象服务状态

```
# swift stat
```

如图12-22所示。

```
[root@controller swift]# swift stat
         Account: AUTH_cec93d07f74a4346a3b2221241abb52e
      Containers: 0
         Objects: 0
           Bytes: 0
  X-Put-Timestamp: 1490621669.02023
     X-Timestamp: 1490621669.02023
       X-Trans-Id: tx7581d8a794e94c109ed54-0058d914e4
    Content-Type: text/plain; charset=utf-8
```

图 12 - 22　列出对象服务状态

3) 创建 container1 容器

\# openstack container create container1

如图 12 - 23 所示。

```
[root@controller swift]# openstack container create container1
+--------------------------------------+------------+------------------------------------+
| account                              | container  | x-trans-id                         |
+--------------------------------------+------------+------------------------------------+
| AUTH_cec93d07f74a4346a3b2221241abb52e | container1 | tx4ef4bc79813a429db9b7a-0058d91509 |
+--------------------------------------+------------+------------------------------------+
```

图 12 - 23　创建 container1 容器

4) 创建一个测试文件，并将其上传到 container1 容器中

\# echo 'this is object storage test file' > /tmp/test.txt

\# openstack object create container1 /tmp/test.txt

如图 12 - 24 所示。

```
[root@controller swift]# echo 'this is object storage test file' > /tmp/test.txt
[root@controller swift]#
[root@controller swift]# openstack object create container1 /tmp/test.txt
+---------------+------------+----------------------------------+
| object        | container  | etag                             |
+---------------+------------+----------------------------------+
| /tmp/test.txt | container1 | ffa9c175f935397383973e1d8d803b1f |
+---------------+------------+----------------------------------+
```

图 12 - 24　上传测试文件

5) 列出 container1 容器里的所有文件

\# openstack object list container1

如图 12 - 25 所示。

```
[root@controller swift]# openstack object list container1
+---------------+
| Name          |
+---------------+
| /tmp/test.txt |
+---------------+
```

图 12 - 25　列出 container1 容器里的所有文件

6）删除本地的删除文件，从 container1 容器里下载一个测试文件，并验证其内容

rm-rf /tmp/test.txt
openstack object save container1 /tmp/test.txt
cat /tmp/test.txt

如图 12-26 所示。

```
[root@controller swift]# rm -rf /tmp/test.txt
[root@controller swift]#
[root@controller swift]# openstack object save container1 /tmp/test.txt
[root@controller swift]#
[root@controller swift]# cat /tmp/test.txt
this is object storage test file
```

图 12-26　验证测试文件

第 13 章　编排（Heat）服务

请在控制节点上完成本章的操作。

13.1　服务概述

　　Heat 服务是指基于模板来编排资源，它通过运行调用生成运行中云应用程序的 OpenStack API 为描述云应用程序提供基于模板的编排。该软件将其他 OpenStack 核心组件整合进一个单文件模板系统。简化了复杂的基础设施、服务和应用的定义和部署，模板允许创建很多种类的 OpenStack 资源，如实例、浮点 IP、云硬盘、安全组和用户等。它也提供高级功能，如实例高可用、实例自动缩放和嵌套栈等。这使得 OpenStack 的核心项目有着庞大的用户群。

　　Heat 服务使部署人员能够直接或者通过定制化插件来与编排服务集成。

编排服务包含以下组件：

1. heat 命令行客户端

　　一个命令行工具，和 heat-api 通信，以运行 AWS CloudFormation API，最终开发者可以直接使用 Orchestration REST API。

2. heat-api 组件

　　一个 OpenStack 本地 REST API，发送 API 请求到 heat-engine，通过远程过程调用（RPC）。

3. heat-api-cfn 组件

　　提供与 AWS CloudFormation 兼容的队列 API，发送 API 请求到 heat-engine，通过远程过程调用（RPC）。

4. heat-engine

　　提供协作功能，启动模板和提供给 API 消费者回馈事件。

13.2　安装前准备（控制节点）

　　安装和配置编排服务之前，必须创建一个数据库、服务凭证和 API 端点。

13.2.1　创建数据库

　　1）用数据库连接客户端以 root 用户连接到数据库服务器

```
# mysql-u root-p'SQL_PASS'
```

2）创建 heat 数据库

```
CREATE DATABASE heat;
```

3）对 heat 数据库授予恰当的权限

本书示例中将 heat 数据库密码设置为 HEAT_DBPASS。

```
GRANT ALL PRIVILEGES ON heat.* TO 'heat'@'localhost' \
    IDENTIFIED BY 'HEAT_DBPASS';
GRANT ALL PRIVILEGES ON heat.* TO 'heat'@'%' \
    IDENTIFIED BY 'HEAT_DBPASS';
```

4）退出数据库客户端

```
exit
```

13.2.2 创建用户、服务实体和 API 端点

1. 获得 admin 凭证

获得 admin 凭证来获取只有管理员才能执行的命令的访问权限：

```
# . /root/admin-openrc
```

2. 创建 heat 用户

1）创建 heat 用户

```
# openstack user create --domain default \
    --password HEAT_PASS heat
```

本书示例中将 heat 用户密码设置为 HEAT_PASS，如图 13-1 所示。

```
[root@controller ~]# openstack user create --domain default \
> --password HEAT_PASS heat
+-----------+----------------------------------+
| Field     | Value                            |
+-----------+----------------------------------+
| domain_id | 1224d01d12d5451191c2f0aeefa6ad6a |
| enabled   | True                             |
| id        | c1558cb4399f448587c3502d6ef93cee |
| name      | heat                             |
+-----------+----------------------------------+
```

图 13-1 创建 heat 用户

2）添加 admin 角色到 heat 用户和 service 项目上

```
# openstack role add --project service --user heat admin
```

如图 13-2 所示。

```
[root@controller ~]# openstack role add --project service --user heat admin
[root@controller ~]#
```

图 13-2 添加 admin 角色到 heat 用户和 service 项目上

3. 创建服务实体和身份认证服务

1）创建 heat 和 heat-cfn 服务实体：

编排服务要求两个服务实体，如图 13-3 和 13-4 所示。

\# openstack service create--name heat \
 --description "Orchestration" orchestration

```
[root@controller ~]# openstack service create --name heat \
> --description "Orchestration" orchestration
+-------------+----------------------------------+
| Field       | Value                            |
+-------------+----------------------------------+
| description | Orchestration                    |
| enabled     | True                             |
| id          | dfd6b3ad6ae944469760ccb10916844e |
| name        | heat                             |
| type        | orchestration                    |
+-------------+----------------------------------+
```

图 13-3　创建 heat 服务实体

\# openstack service create--name heat-cfn \
 --description "Orchestration" cloudformation

```
[root@controller ~]# openstack service create --name heat-cfn \
> --description "Orchestration" cloudformation
+-------------+----------------------------------+
| Field       | Value                            |
+-------------+----------------------------------+
| description | Orchestration                    |
| enabled     | True                             |
| id          | 0f85d9d3fc6c47209cbd8d543a5cba63 |
| name        | heat-cfn                         |
| type        | cloudformation                   |
+-------------+----------------------------------+
```

图 13-4　创建 heat-cfn 服务实体

4. 创建 API 端点

创建编排服务 API 端点：

编排服务每个服务实体都需要端点，如图 13-5 至 13-10 所示。

\# openstack endpoint create--region RegionOne \
orchestration public http：//controller：8004/v1/% \（tenant_id \）s

```
[root@controller ~]# openstack endpoint create --region RegionOne \
> orchestration public http://controller:8004/v1/%\(tenant_id\)s
+--------------+----------------------------------------------+
| Field        | Value                                        |
+--------------+----------------------------------------------+
| enabled      | True                                         |
| id           | 8baa30927fed42bd917790e13b481b4f             |
| interface    | public                                       |
| region       | RegionOne                                    |
| region_id    | RegionOne                                    |
| service_id   | dfd6b3ad6ae944469760ccb10916844e             |
| service_name | heat                                         |
| service_type | orchestration                                |
| url          | http://controller:8004/v1/%(tenant_id)s      |
+--------------+----------------------------------------------+
```

图 13 - 5　创建编排服务 heat API 端点（public）

\# openstack endpoint create--region RegionOne \

　　orchestration internal http：//controller：8004/v1/% \ （tenant_id \) s

```
[root@controller ~]# openstack endpoint create --region RegionOne \
> orchestration internal http://controller:8004/v1/%\(tenant_id\)s
+--------------+----------------------------------------------+
| Field        | Value                                        |
+--------------+----------------------------------------------+
| enabled      | True                                         |
| id           | eeb98af6bfea4efab581d5b14369c040             |
| interface    | internal                                     |
| region       | RegionOne                                    |
| region_id    | RegionOne                                    |
| service_id   | dfd6b3ad6ae944469760ccb10916844e             |
| service_name | heat                                         |
| service_type | orchestration                                |
| url          | http://controller:8004/v1/%(tenant_id)s      |
+--------------+----------------------------------------------+
```

图 13 - 6　创建编排服务 heat API 端点（internal）

\# openstack endpoint create--region RegionOne \

　　orchestration admin http：//controller：8004/v1/% \ （tenant_id \) s

```
[root@controller ~]# openstack endpoint create --region RegionOne \
> orchestration admin http://controller:8004/v1/%\(tenant_id\)s
+--------------+----------------------------------------------+
| Field        | Value                                        |
+--------------+----------------------------------------------+
| enabled      | True                                         |
| id           | 01301baf004747e7a0e4f2d891486062             |
| interface    | admin                                        |
| region       | RegionOne                                    |
| region_id    | RegionOne                                    |
| service_id   | dfd6b3ad6ae944469760ccb10916844e             |
| service_name | heat                                         |
| service_type | orchestration                                |
| url          | http://controller:8004/v1/%(tenant_id)s      |
+--------------+----------------------------------------------+
```

图 13 - 7　创建编排服务 heat API 端点（adimn）

openstack endpoint create--region RegionOne \
　　cloudformation public http：//controller：8000/v1

```
[root@controller ~]# openstack endpoint create --region RegionOne \
> cloudformation public http://controller:8000/v1
+--------------+----------------------------------+
| Field        | Value                            |
+--------------+----------------------------------+
| enabled      | True                             |
| id           | 1548d2c7829d43ea8483420d122e8253 |
| interface    | public                           |
| region       | RegionOne                        |
| region_id    | RegionOne                        |
| service_id   | 0f85d9d3fc6c47209cbd8d543a5cba63 |
| service_name | heat-cfn                         |
| service_type | cloudformation                   |
| url          | http://controller:8000/v1        |
+--------------+----------------------------------+
```

图 13-8　创建编排服务 heat-cfn API 端点（public）

openstack endpoint create--region RegionOne \
　　cloudformation internal http：//controller：8000/v1

```
[root@controller ~]# openstack endpoint create --region RegionOne \
> cloudformation internal http://controller:8000/v1
+--------------+----------------------------------+
| Field        | Value                            |
+--------------+----------------------------------+
| enabled      | True                             |
| id           | d145db925812411a93eab430993e29cc |
| interface    | internal                         |
| region       | RegionOne                        |
| region_id    | RegionOne                        |
| service_id   | 0f85d9d3fc6c47209cbd8d543a5cba63 |
| service_name | heat-cfn                         |
| service_type | cloudformation                   |
| url          | http://controller:8000/v1        |
+--------------+----------------------------------+
```

图 13-9　创建编排服务 heat-cfn API 端点（internal）

openstack endpoint create--region RegionOne \
　　cloudformation admin http：//controller：8000/v1

```
[root@controller ~]# openstack endpoint create --region RegionOne \
> cloudformation admin http://controller:8000/v1
+--------------+----------------------------------+
| Field        | Value                            |
+--------------+----------------------------------+
| enabled      | True                             |
| id           | 8cafd23100634cc8a1d86e36f830ec34 |
| interface    | admin                            |
| region       | RegionOne                        |
| region_id    | RegionOne                        |
| service_id   | 0f85d9d3fc6c47209cbd8d543a5cba63 |
| service_name | heat-cfn                         |
| service_type | cloudformation                   |
| url          | http://controller:8000/v1        |
+--------------+----------------------------------+
```

图 13-10　创建编排服务 heat-cfn API 端点（admin）

13.2.3　创建域、项目、用户和角色

为了管理栈，在认证服务中编排服务需要更多信息。

1）为栈创建 heat 包含项目和用户的域

\# openstack domain create--description "Stack projects and users" heat

如图 13-11 所示。

```
[root@controller ~]# openstack domain create --description "Stack projects and users" heat
+-------------+----------------------------------+
| Field       | Value                            |
+-------------+----------------------------------+
| description | Stack projects and users         |
| enabled     | True                             |
| id          | b91fa5658cd9494ba23f10a1c9001138 |
| name        | heat                             |
+-------------+----------------------------------+
```

图 13-11　为栈创建 heat 包含项目和用户的域

2）在 heat 域中创建管理项目和用户的 heat_domain_admin 用户

\# openstack user create--domain heat--password HEAT_DOMAIN_PASS \
　heat_domain_admin

本书示例中将 heat_domain_admin 用户密码设置为 HEAT_DOMAIN_PASS，如图 13-12 所示。

```
[root@controller ~]# openstack user create --domain heat --password HEAT_DOMAIN_PASS \
> heat_domain_admin
+-----------+----------------------------------+
| Field     | Value                            |
+-----------+----------------------------------+
| domain_id | b91fa5658cd9494ba23f10a1c9001138 |
| enabled   | True                             |
| id        | 48be91277bb44972b4068eed21fcf074 |
| name      | heat_domain_admin                |
+-----------+----------------------------------+
```

图 13-12　在 heat 域中创建管理项目和用户的 heat_domain_admin 用户

3）添加 admin 角色到 heat 域中的 heat_domain_admin 用户，启用 heat_domain_admin 用户管理栈的管理权限

\# openstack role add--domain heat--user-domain heat \
   --user heat_domain_admin admin

如图 13-13 所示。

```
[root@controller ~]# openstack role add --domain heat --user-domain heat \
> --user heat_domain_admin admin
[root@controller ~]#
```

图 13-13　添加 admin 角色到 heat 域中的 heat_domain_admin 用户，启用 heat_domain_admin 用户管理栈的管理权限

4）创建 heat_stack_owner 角色

\# openstack role create heat_stack_owner

如图 13-14 所示。

```
[root@controller ~]# openstack role create heat_stack_owner
+-----------+----------------------------------+
| Field     | Value                            |
+-----------+----------------------------------+
| domain_id | None                             |
| id        | 133235eb58b9498f82c8a3308a4f8375 |
| name      | heat_stack_owner                 |
+-----------+----------------------------------+
```

图 13-14　创建 heat_stack_owner 角色

5）添加 heat_stack_owner 角色到 demo 项目和用户，启用 demo 用户管理栈

\# openstack role add--project demo--user demo heat_stack_owner

如图 13-15 所示。

```
[root@controller ~]# openstack role add --project demo --user demo heat_stack_owner
[root@controller ~]#
```

图 13-15　添加 heat_stack_owner 角色到 demo 项目和用户，启用 demo 用户管理栈

必须添加 heat_stack_owner 角色到每个管理栈的用户。

6）创建 heat_stack_user 角色

\# openstack role create heat_stack_user

如图 13-16 所示。

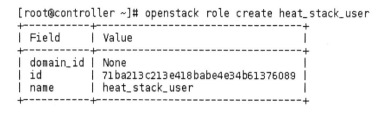

图 13-16　创建 heat_stack_user 角色

编排服务自动地分配 heat_stack_user 角色给在 stack 部署过程中创建的用户。默认情况下，这个角色会限制 API 的操作。为了避免冲突，请不要为用户添加 heat_stack_owner 角色。

13.3　安装和配置（控制节点）

13.3.1　安装软件包

安装 glance 软件包：
yum-y install openstack-heat-api openstack-heat-api-cfn \
　openstack-heat-engine

13.3.2　配置 Heat

编辑 /etc/heat/heat.conf 文件：
由于默认配置文件在各发行版本中可能不同，因此，在进行修改的同时可能需要添加部分选项。另外，在配置片段中的省略号（...）表示默认的配置选项，应该保留。
gedit /etc/heat/heat.conf
a. 在 [database] 部分，配置数据库访问：

[database]
...
connection = mysql + pymysql: //heat: HEAT_DBPASS@ controller/heat

b. 在 [DEFAULT] 和 [oslo_messaging_rabbit] 部分，配置 RabbitMQ 消息队列访问：

[DEFAULT]
...
rpc_backend = rabbit

[oslo_messaging_rabbit]
...
rabbit_host = controller
rabbit_userid = openstack
rabbit_password = RABBIT_PASS

c. 在 [keystone_authtoken]、[trustee]、[clients_keystone] 和 [ec2authtoken] 部分，配置认证服务访问：

由于初始配置文件中并没有 [keystone_authtoken] 部分，因此，可以在文件尾部添加此部分内容。

[clients_keystone]
...
auth_uri = http://controller:35357

[ec2authtoken]
...
auth_uri = http://controller:5000/v2.0

[trustee]
...
auth_plugin = password
auth_url = http://controller:35357
username = heat
user_domain_name = default
password = HEAT_PASS

[keystone_authtoken]
auth_uri = http://controller:5000
auth_url = http://controller:35357
memcached_servers = controller:11211
auth_type = password
project_domain_name = default
user_domain_name = default
project_name = service
username = heat
password = HEAT_PASS

d. 在［DEFAULT］部分，配置元数据和等待条件 URLs：

［DEFAULT］
…
heat_metadata_server_url = http：//controller：8000
heat_waitcondition_server_url = http：//controller：8000/v1/waitcondition

e. 在［DEFAULT］部分，配置栈域与管理凭据：

［DEFAULT］
…
stack_user_domain_name = heat
stack_domain_admin = heat_domain_admin
stack_domain_admin_password = HEAT_DOMAIN_PASS

13.3.3 写入数据库

写入编排服务数据库：

su-s /bin/sh-c "heat-manage db_sync" heat

如图 13-17 所示。

```
[root@controller ~]# su -s /bin/sh -c "heat-manage db_sync" heat
2017-03-28 09:10:25.022 5115 INFO migrate.versioning.api [-] 27 -> 28...
2017-03-28 09:10:25.285 5115 INFO migrate.versioning.api [-] done
2017-03-28 09:10:25.285 5115 INFO migrate.versioning.api [-] 28 -> 29...
2017-03-28 09:10:25.374 5115 INFO migrate.versioning.api [-] done
2017-03-28 09:10:25.374 5115 INFO migrate.versioning.api [-] 29 -> 30...
2017-03-28 09:10:25.395 5115 INFO migrate.versioning.api [-] done
2017-03-28 09:10:25.395 5115 INFO migrate.versioning.api [-] 30 -> 31...
2017-03-28 09:10:25.426 5115 INFO migrate.versioning.api [-] done
```

图 13-17 写入编排服务数据库

忽略输出中任何不推荐使用的信息。

13.3.4 完成安装

启动编排服务，并将其配置为随系统启动：

systemctl enable openstack-heat-api.service \
    openstack-heat-api-cfn.service openstack-heat-engine.service
systemctl start openstack-heat-api.service \
    openstack-heat-api-cfn.service openstack-heat-engine.service

13.4　验证操作（控制节点）

1）获得 admin 凭证来获取只有管理员才能执行的命令的访问权限

\# ./root/admin-openrc

2）列出服务组件，以验证是否成功启动并注册了每个进程

\# openstack orchestration service list

图 13-18 所示。

```
[root@controller ~]# openstack orchestration service list
+------------+-------------+-------------------------+------------+--------+----------------------+--------+
| hostname   | binary      | engine_id               | host       | topic  | updated_at           | status |
+------------+-------------+-------------------------+------------+--------+----------------------+--------+
| controller | heat-engine | 8a63291a-8340-4789-a47f- | controller | engine | 2017-03-28T01:11:    | up     |
|            |             | 40c67be4449e            |            |        | 13.000000            |        |
| controller | heat-engine | 38895807-fe14-47c4-815c-| controller | engine | 2017-03-28T01:11:    | up     |
|            |             | 998a63cf9300            |            |        | 13.000000            |        |
| controller | heat-engine | e96d912e-a804-4898-a3d8-| controller | engine | 2017-03-28T01:11:    | up     |
|            |             | f83782ff8864            |            |        | 13.000000            |        |
| controller | heat-engine | 9ec0e31a-7dcf-4240-     | controller | engine | 2017-03-28T01:11:    | up     |
|            |             | 9da5-1a0c44db7595       |            |        | 13.000000            |        |
+------------+-------------+-------------------------+------------+--------+----------------------+--------+
```

图 13-18　验证注册

在正确配置的情况下，控制节点上应该输出 heat-engine 组件。

第 14 章 计量（Ceilometer）服务

请在控制节点和相关节点上完成本章的操作。

14.1 服务概述

14.1.1 计量数据收集服务提供的功能

（1）相关 OpenStack 服务的有效调查计量数据；
（2）通过监测通知收集来自各个服务发送的事件和计量数据；
（3）发布收集来的数据到多个目标，包括数据存储和消息队列。

14.1.2 计量数据收集服务包含的组件

1．ceilometer-agent-compute（计算代理）
运行在每个计算节点中，以推送资源的使用状态，目前主要专注于创建计算节点代理。

2．ceilometer-agent-central（中心代理）
运行在中心管理服务器中，以推送资源使用状态。它无须捆绑到虚拟机实例，也无须运行在计算节点服务器上。此代理服务可以进行横向扩展。

3．ceilometer（通知代理）
运行在中心管理服务器中，通过获取来自消息队列的消息，再构建事件和计量数据。

4．ceilometor 收集器（负责接收信息进行持久化存储）
运行在中心管理服务器中，将收集到的原始计量数据分发到数据存储或者外部使用者。

5．ceilometer-api（API 服务器）
运行在一个或多个中心管理服务器，提供数据存储的数据访问。

14.1.3 计量报警服务提供的功能

（1）当收集的度量或事件数据打破了界定的规则时，计量报警服务会触发报警；
（2）这些服务使用 OpenStack 消息总线来通信，只有收集者和 API 服务才可以访问数据存储。

14.1.4 计量报警服务包含的组件

1．aodh-api（API 服务器）

运行于一个或多个中心管理服务器中，提供访问存储在数据中心的警告信息的服务。

2．aodh-evaluator（报警评估器）

运行在一个或多个中心管理服务器中，当与警告相关联的统计趋势超过指定的时间窗口阈值时，将会做出相应的决定。

3．aodh-listener（通知监听器）

运行在一个中心管理服务器中，来检测什么时候发出告警。根据一些事件预先定义的规则，产生相应的告警，同时能够被数据收集服务的通知代理捕获到。

4．aodh-notifier（报警通知器）

运行在一个或多个中心管理服务器中，允许警告为一组收集的实例基于评估阀值来设置。

这些服务使用 OpenStack 消息总线来通信，只有收集者和 API 服务才可以访问数据存储。

14.2 安装前准备（控制节点）

安装和配置计量服务之前，必须创建一个数据库、服务凭证和 API 端点。

14.2.1 安装和配置 NoSQL 数据库

计量服务使用 NoSQL 数据库来存储信息，NoSQL 数据库一般运行在控制节点上。本书示例中使用的是 MongoDB 数据库。

1）安装 MongoDB 软件包

yum -y install mongodb-server mongodb

2）编辑 /etc/mongod.conf 文件

gedit /etc/mongod.conf

a．配置 bind_ip 使用控制节点管理网卡的 IP 地址：

bind_ip = MANAGEMENT_INTERFACE_IP_ADDRESS

将 MANAGEMENT_INTERFACE_IP_ADDRESS 替换为控制节点的管理网络 IP 地址，本书示例中将其替换为 192.168.1.11。

b．配置 smallfiles 值，是否限制日志大小：

默认情况下，MongoDB 会在 /var/lib/mongodb/journal 目录下创建几个 1GB 大小的日

志文件。

如果想将每个日志文件大小减小到 128MB 并且限制日志文件占用的总空间为 512MB，则需将 smallfiles 值设置为 True：

smallfiles = true

3）启动 MongoDB 服务，并将其配置为随系统启动

\# systemctl enable mongod.service

\# systemctl start mongod.service

14.2.2 创建数据库

创建 ceilometer 数据库：

\# mongo--host controller--eval '
   db = db.getSiblingDB（"ceilometer"）;
   db.createUser（｛user：" ceilometer"，
   pwd："CEILOMETER_DBPASS"，
   roles：["readWrite"，"dbAdmin"]｝）'

本书示例中将 ceilometer 数据库密码设置为 CEILOMETER_DBPASS，如图 14-1 所示。

```
[root@controller ~]# mongo --host controller --eval '
> db = db.getSiblingDB("ceilometer");
> db.createUser({user: "ceilometer",
> pwd: "CEILOMETER_DBPASS",
> roles: [ "readWrite", "dbAdmin" ]})'
MongoDB shell version: 2.6.11
connecting to: controller:27017/test
Successfully added user: { "user" : "ceilometer", "roles" : [ "readWrite", "dbAdmin" ] }
```

图 14-1 创建 ceilometer 数据库

14.2.3 创建用户、服务实体和 API 端点

1. 获得 admin 凭证

获得 admin 凭证来获取只有管理员才能执行的命令的访问权限：

\# ./root/admin-openrc

2. 创建 ceilometer 用户

1）创建 ceilometer 用户

\# openstack user create--domain default \
   --password CEILOMETER_PASS ceilometer

本书示例中将 ceilometer 用户密码设置为 CEILOMETER_PASS，如图 14-2 所示。

```
[root@controller ~]# openstack user create --domain default \
> --password CEILOMETER_PASS ceilometer
+-----------+----------------------------------+
| Field     | Value                            |
+-----------+----------------------------------+
| domain_id | 1224d01d12d5451191c2f0aeefa6ad6a |
| enabled   | True                             |
| id        | b79de095a475484988287ce675ceb3df |
| name      | ceilometer                       |
+-----------+----------------------------------+
```

图 14-2 创建 ceilometer 用户

2）添加 admin 角色到 ceilometer 用户和 service 项目上

openstack role add--project service--user ceilometer admin

如图 14-3 所示。

```
[root@controller ~]# openstack role add --project service --user ceilometer admin
[root@controller ~]#
```

图 14-3 添加 admin 角色到 ceilometer 用户和 service 项目上

3. 创建服务实体和身份认证服务

创建 ceilometer 服务实体：

openstack service create--name heat \
 --description "Telemetry" metering

如图 14-4 所示。

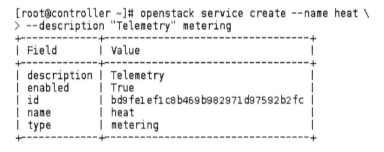

图 14-4 创建 ceilometer 服务实体

4. 创建 API 端点

创建计量服务 API 端点，如图 14-5 到 14-7 所示。

openstack endpoint create--region RegionOne \
 metering public http：//controller：8777

```
[root@controller ~]# openstack endpoint create --region RegionOne \
> metering public http://controller:8777
+--------------+----------------------------------+
| Field        | Value                            |
+--------------+----------------------------------+
| enabled      | True                             |
| id           | e472c2fe2ba24b13b502c25791236c3e |
| interface    | public                           |
| region       | RegionOne                        |
| region_id    | RegionOne                        |
| service_id   | bd9fe1ef1c8b469b982971d97592b2fc |
| service_name | heat                             |
| service_type | metering                         |
| url          | http://controller:8777           |
+--------------+----------------------------------+
```

图 14-5　创建计量服务 API 端点（public）

openstack endpoint create--region RegionOne \

　metering internal http：//controller：8777

```
[root@controller ~]# openstack endpoint create --region RegionOne \
> metering internal http://controller:8777
+--------------+----------------------------------+
| Field        | Value                            |
+--------------+----------------------------------+
| enabled      | True                             |
| id           | b841ca852ac24a67ab7c9955f67604b5 |
| interface    | internal                         |
| region       | RegionOne                        |
| region_id    | RegionOne                        |
| service_id   | bd9fe1ef1c8b469b982971d97592b2fc |
| service_name | heat                             |
| service_type | metering                         |
| url          | http://controller:8777           |
+--------------+----------------------------------+
```

图 14-6　创建计量服务 API 端点（internal）

openstack endpoint create--region RegionOne \

　metering admin http：//controller：8777

```
[root@controller ~]# openstack endpoint create --region RegionOne \
> metering admin http://controller:8777
+--------------+----------------------------------+
| Field        | Value                            |
+--------------+----------------------------------+
| enabled      | True                             |
| id           | e44144c6b701474e8735f630556d68aa |
| interface    | admin                            |
| region       | RegionOne                        |
| region_id    | RegionOne                        |
| service_id   | bd9fe1ef1c8b469b982971d97592b2fc |
| service_name | heat                             |
| service_type | metering                         |
| url          | http://controller:8777           |
+--------------+----------------------------------+
```

图 14-7　创建计量服务 API 端点（admin）

14.3 安装和配置（控制节点）

14.3.1 安装软件包

安装 ceilometer 软件包：

```
# yum -y install openstack-ceilometer-api \
    openstack-ceilometer-collector openstack-ceilometer-notification \
    openstack-ceilometer-central python-ceilometerclient
```

14.3.2 配置 Ceilometer

编辑 /etc/ceilometer/ceilometer.conf 文件：

由于默认配置文件在各发行版本中可能不同，因此，在进行修改的同时可能需要添加部分选项。另外，在配置片段中的省略号（...）表示默认的配置选项，应该保留。

```
# gedit /etc/ceilometer/ceilometer.conf
```

a. 在 [database] 部分，配置数据库访问：

```
[database]
...
connection = mongodb://ceilometer:CEILOMETER_DBPASS@controller:27017/ceilometer
```

b. 在 [DEFAULT] 和 [oslo_messaging_rabbit] 部分，配置 RabbitMQ 消息队列访问：

```
[DEFAULT]
...
rpc_backend = rabbit
[oslo_messaging_rabbit]
...
rabbit_host = controller
rabbit_userid = openstack
rabbit_password = RABBIT_PASS
```

c. 在 [DEFAULT] 和 [keystone_authtoken] 部分，配置认证服务访问：

```
[DEFAULT]
...
auth_strategy = keystone

[keystone_authtoken]
...
auth_uri = http://controller:5000
auth_url = http://controller:35357
memcached_servers = controller:11211
auth_type = password
project_domain_name = default
user_domain_name = default
project_name = service
username = ceilometer
password = CEILOMETER_PASS
```

d. 在[service_credentials]部分，配置服务证书：

```
[service_credentials]
...
region_name = RegionOne
interface = internalURL
auth_type = password
auth_url = http://controller:5000/v3
project_name = service
project_domain_name = default
username = ceilometer
user_domain_name = default
password = CEILOMETER_PASS
```

14.3.3 完成安装

启动计量服务，并将其配置为随系统启动：
```
# systemctl enable openstack-ceilometer-api.service \
    openstack-ceilometer-notification.service \
    openstack-ceilometer-central.service \
```

openstack-ceilometer-collector.service

systemctl start openstack-ceilometer-api.service \

openstack-ceilometer-notification.service \

openstack-ceilometer-central.service \

openstack-ceilometer-collector.service

14.4 启用镜像服务计量（控制节点）

Telemetry 使用通知收集镜像服务计量信息。

14.4.1 配置镜像服务使用计量服务

1）编辑 /etc/glance/glance-api.conf 文件

由于默认配置文件在各发行版本中可能不同，因此，在进行修改的同时可能需要添加部分选项。另外，在配置片段中的省略号（…）表示默认的配置选项，应该保留。

gedit /etc/glance/glance-api.conf

在［DEFAULT］、［oslo_messaging_notifications］和［oslo_messaging_rabbit］部分，配置通知和 RabbitMQ 消息队列的连接：

```
[DEFAULT]
…
rpc_backend = rabbit

[oslo_messaging_notifications]
…
driver = messagingv2

[oslo_messaging_rabbit]
…
rabbit_host = controller
rabbit_userid = openstack
rabbit_password = RABBIT_PASS
```

2）编辑 /etc/glance/glance-registry.conf 文件

由于默认配置文件在各发行版本中可能不同，因此，在进行修改的同时可能需要添加部分选项。另外，在配置片段中的省略号（…）表示默认的配置选项，应该保留。

gedit /etc/glance/glance-registry.conf

在［DEFAULT］、［oslo_messaging_notifications］和［oslo_messaging_rabbit］部分，配置通知和RabbitMQ消息队列的连接：

```
[DEFAULT]
...
rpc_backend = rabbit

[oslo_messaging_notifications]
...
driver = messagingv2

[oslo_messaging_rabbit]
...
rabbit_host = controller
rabbit_userid = openstack
rabbit_password = RABBIT_PASS
```

14.4.2 完成安装

重启镜像服务：

systemctl restart openstack-glance-api.service \
　　openstack-glance-registry.service

14.5 启用计算服务计量（计算节点）

计量服务通过结合使用通知和代理来收集Computer度量值。在每个计算节点上执行这些步骤。

14.5.1 安装和配置

1）安装ceilometer软件包

yum -y install openstack-ceilometer-compute \
　　python-ceilometerclient python-pecan

2）编辑/etc/ceilometer/ceilometer.conf文件

由于默认配置文件在各发行版本中可能不同，因此，在进行修改的同时可能需要添加部分选项。另外，在配置片段中的省略号（...）表示默认的配置选项，应该保留。

gedit /etc/ceilometer/ceilometer.conf

a. 在［DEFAULT］和［oslo_messaging_rabbit］部分，配置 RabbitMQ 消息队列访问：

［DEFAULT］
...
rpc_backend = rabbit

［oslo_messaging_rabbit］
...
rabbit_host = controller
rabbit_userid = openstack
rabbit_password = RABBIT_PASS

b. 在［DEFAULT］和［keystone_authtoken］部分，配置认证服务访问：

［DEFAULT］
...
auth_strategy = keystone

［keystone_authtoken］
...
auth_uri = http：//controller：5000
auth_url = http：//controller：35357
memcached_servers = controller：11211
auth_type = password
project_domain_name = default
user_domain_name = default
project_name = service
username = ceilometer
password = CEILOMETER_PASS

c. 在［service_credentials］部分，配置服务证书：

［service_credentials］
...
region_name = RegionOne
interface = internalURL
auth_type = password

```
auth_url = http://controller:5000/v3
project_name = service
project_domain_name = default
username = ceilometer
user_domain_name = default
password = CEILOMETER_PASS
```

14.5.2 配置计算服务使用计量服务

编辑 /etc/nova/nova.conf 文件：

由于默认配置文件在各发行版本中可能不同，因此，在进行修改的同时可能需要添加部分选项。另外，在配置片段中的省略号（...）表示默认的配置选项，应该保留。

```
# gedit /etc/nova/nova.conf
```

在 [DEFAULT] 部分，配置提醒：

```
[DEFAULT]
...
notify_on_state_change = vm_and_task_state
notification_driver = messagingv2
instance_usage_audit_period = hour
instance_usage_audit = True
```

14.5.3 完成安装

1) 当系统启动时，启动计量服务并配置它启动

```
# systemctl enable openstack-ceilometer-compute.service
# systemctl start openstack-ceilometer-compute.service
```

2) 重启计算服务

```
# systemctl restart openstack-nova-compute.service
```

14.6 启用块存储计量（控制节点、块存储节点）

计量服务使用通知收集块存储服务计量。在控制节点和块存储节点上执行这些步骤。

14.6.1 配置卷使用计量服务

在控制节点和存储节点上，编辑 /etc/cinder/cinder.conf 文件：

由于默认配置文件在各发行版本中可能不同，因此，在进行修改的同时可能需要添加部分选项。另外，在配置片段中的省略号（…）表示默认的配置选项，应该保留。

gedit /etc/cinder/cinder.conf

在［oslo_messaging_notifications］部分，配置提醒：

［oslo_messaging_notifications］
…
driver = messagingv2

14.6.2 完成安装

1）重启控制节点上的块存储服务

systemctl restart openstack-cinder-api.service \
    openstack-cinder-scheduler.service

2）重启存储节点上的块存储卷服务

systemctl restart openstack-cinder-volume.service

14.7 启用对象存储服务计量（控制节点）

计量服务要求用 ResellerAdmin 的角色来访问对象存储服务。在控制节点上实施这些步骤。

14.7.1 安装前准备

1）获得 admin 凭证来获取只有管理员才能执行的命令的访问权限

./root/admin-openrc

2）创建 ResellerAdmin 角色：

openstack role create ResellerAdmin

如图 14-8 所示。

图 14-8 创建 ResellerAdmin 角色

3) 添加 ResellerAdmin 角色到 ceilometer 用户上

openstack role add--project service--user ceilometer ResellerAdmin

如图 14-9 所示。

```
[root@controller ~]# openstack role add --project service --user ceilometer ResellerAdmin
[root@controller ~]#
```

图 14-9 添加 ResellerAdmin 角色到 ceilometer 用户上

14.7.2 安装组件

安装软件包：

yum-y install python-ceilometermiddleware

14.7.3 配置对象存储服务使用计量服务

在控制节点和其他运行了对象存储的代理服务的节点上执行这些步骤。

编辑 /etc/swift/proxy-server.conf 文件：

由于默认配置文件在各发行版中可能不同，因此，在进行修改的同时可能需要添加部分选项。另外，在配置片段中的省略号（...）表示默认的配置选项，应该保留。

gedit /etc/swift/proxy-server.conf

a. 在 [filter：keystoneauth] 部分，添加 ResellerAdmin 角色：

[filter：keystoneauth]
...
operator_roles = admin, user, ResellerAdmin

b. 在 [pipeline：main] 部分，添加 ceilometer 用户：

[pipeline：main]
...
pipeline = ceilometer catch_errors gatekeeper healthcheck proxy-logging cache container_sync bulk ratelimit authtoken keystoneauth container-quotas account-quotas slo dlo versioned_writes proxy-logging proxy-server

c. 在 [filter：ceilometer] 部分，配置提醒：

由于初始配置文件中并没有 [filter：ceilometer] 部分，因此可以在文件尾部添加此部分内容。

```
[filter：ceilometer]
paste.filter_factory = ceilometermiddleware.swift：filter_factory
control_exchange = swift
url = rabbit：//openstack：RABBIT_PASS@controller：5672/
driver = messagingv2
topic = notifications
log_level = WARN
```

14.7.4 完成安装

在控制节点和其他运行了对象存储的代理服务的节点上重启对象存储的代理服务。
重启对象存储的代理服务：

```
# systemctl restart openstack-swift-proxy.service
```

14.8 安装计量警告服务（控制节点）

安装和配置网络服务之前，必须创建一个数据库、服务凭证和 API 端点。

14.8.1 创建数据库

1）用数据库连接客户端以 root 用户连接到数据库服务器
```
# mysql-u root-p'SQL_PASS'
```
2）创建 aodh 数据库
```
CREATE DATABASE aodh;
```
3）对 aodh 数据库授予恰当的权限
本书示例中将 aodh 数据库密码设置为 AODH_DBPASS。
```
GRANT ALL PRIVILEGES ON aodh.* TO 'aodh'@'localhost' \
  IDENTIFIED BY 'AODH_DBPASS';
GRANT ALL PRIVILEGES ON aodh.* TO 'aodh'@'%' \
  IDENTIFIED BY 'AODH_DBPASS';
```
4）退出数据库客户端
```
exit
```

14.8.2 创建用户、服务实体和 API 端点

1. 获得 admin 凭证

获得 admin 凭证来获取只有管理员才能执行的命令的访问权限：

./root/admin-openrc

2. 创建 aodh 用户

1) 创建 aodh 用户

openstack user create--domain default \
　　--password AODH_PASS aodh

本书示例中将 aodh 用户密码设置为 AODH_PASS，如图 14 – 10 所示。

图 14 – 10　创建 aodh 用户

2) 添加 admin 角色到 aodh 用户和 service 项目上

openstack role add--project service--user aodh admin

如图 14 – 11 所示。

```
[root@controller ~]# openstack role add --project service --user aodh admin
[root@controller ~]#
```

图 14 – 11　添加 admin 角色到 aodh 用户和 service 项目上

3. 创建服务实体和身份认证服务

创建 aodh 服务实体：

openstack service create--name aodh \
　　--description "Telemetry" alarming

如图 14 – 12 所示。

```
[root@controller ~]# openstack service create --name aodh \
> --description "Telemetry" alarming
+-------------+----------------------------------+
| Field       | Value                            |
+-------------+----------------------------------+
| description | Telemetry                        |
| enabled     | True                             |
| id          | 801918330bde4c488c5d131c2b50bfbd |
| name        | aodh                             |
| type        | alarming                         |
+-------------+----------------------------------+
```

图 14 – 12　创建 aodh 服务实体

4. 创建 API 端点

创建网络服务的 API 端点，如图 14-13 到图 14-15 所示。

openstack endpoint create--region RegionOne \

    alarming public http：//controller：8042

```
[root@controller ~]# openstack endpoint create --region RegionOne \
> alarming public http://controller:8042
+--------------+----------------------------------+
| Field        | Value                            |
+--------------+----------------------------------+
| enabled      | True                             |
| id           | c0b9c09a35fb45e8854c5d22abcba097 |
| interface    | public                           |
| region       | RegionOne                        |
| region_id    | RegionOne                        |
| service_id   | 801918330bde4c488c5d131c2b50bfbd |
| service_name | aodh                             |
| service_type | alarming                         |
| url          | http://controller:8042           |
+--------------+----------------------------------+
```

图 14-13　创建网络服务的 API 端点（public）

openstack endpoint create--region RegionOne \

    alarming internal http：//controller：8042

```
[root@controller ~]# openstack endpoint create --region RegionOne \
> alarming internal http://controller:8042
+--------------+----------------------------------+
| Field        | Value                            |
+--------------+----------------------------------+
| enabled      | True                             |
| id           | 55f778715eb9410392f2cddf7ec1befa |
| interface    | internal                         |
| region       | RegionOne                        |
| region_id    | RegionOne                        |
| service_id   | 801918330bde4c488c5d131c2b50bfbd |
| service_name | aodh                             |
| service_type | alarming                         |
| url          | http://controller:8042           |
+--------------+----------------------------------+
```

图 14-14　创建网络服务的 API 端点（internal）

openstack endpoint create--region RegionOne \

    alarming admin http：//controller：8042

```
[root@controller ~]# openstack endpoint create --region RegionOne \
> alarming admin http://controller:8042
+--------------+----------------------------------+
| Field        | Value                            |
+--------------+----------------------------------+
| enabled      | True                             |
| id           | cc9f1da960c743c8a5e4350a2243ee4e |
| interface    | admin                            |
| region       | RegionOne                        |
| region_id    | RegionOne                        |
| service_id   | 801918330bde4c488c5d131c2b50bfbd |
| service_name | aodh                             |
| service_type | alarming                         |
| url          | http://controller:8042           |
+--------------+----------------------------------+
```

图 14-15　创建网络服务的 API 端点（admin）

14.8.3　安装和配置

1) 安装软件包

yum -y install openstack-aodh-api \

　openstack-aodh-evaluator openstack-aodh-notifier \

　openstack-aodh-listener openstack-aodh-expirer \

　python-ceilometerclient

2) 编辑 /etc/aodh/aodh.conf 文件

由于默认配置文件在各发行版本中可能不同，因此，在进行修改的同时可能需要添加部分选项。另外，在配置片段中的省略号（...）表示默认的配置选项，应该保留。

gedit /etc/aodh/aodh.conf

a. 在 [database] 部分，配置数据库访问：

[database]

...

connection = mysql + pymysql：//aodh：AODH_DBPASS@ controller/aodh

b. 在 [DEFAULT] 和 [oslo_messaging_rabbit] 部分，配置 RabbitMQ 消息队列访问：

[DEFAULT]

...

rpc_backend = rabbit

[oslo_messaging_rabbit]

...

rabbit_host = controller

rabbit_userid = openstack

rabbit_password = RABBIT_PASS

c. 在 [DEFAULT] 和 [keystone_authtoken] 部分，配置认证服务访问：

[DEFAULT]
...
auth_strategy = keystone

[keystone_authtoken]
...
auth_uri = http://controller:5000
auth_url = http://controller:35357
memcached_servers = controller:11211
auth_type = password
project_domain_name = default
user_domain_name = default
project_name = service
username = aodh
password = AODH_PASS

d. 在 [service_credentials] 部分，配置服务证书：

[service_credentials]
...
region_name = RegionOne
interface = internalURL
auth_type = password
auth_url = http://controller:5000/v3
project_name = service
project_domain_name = default
username = aodh
user_domain_name = default
password = AODH_PASS

3）初始化告警数据库

su -s /bin/sh -c "aodh-dbsync" aodh

如图 14-16 所示。

```
[root@controller ~]# su -s /bin/sh -c "aodh-dbsync" aodh
[root@controller ~]#
```

图 14-16 初始化告警数据库

14.8.4 完成安装

1）当系统启动时，启动警告服务并配置它启动。

```
# systemctl enable openstack-aodh-api.service \
  openstack-aodh-evaluator.service \
  openstack-aodh-notifier.service \
  openstack-aodh-listener.service
# systemctl start openstack-aodh-api.service \
  openstack-aodh-evaluator.service \
  openstack-aodh-notifier.service \
  openstack-aodh-listener.service
```

14.9 验证操作（控制节点）

1）获得 admin 凭证来获取只有管理员才能执行的命令的访问权限

```
# ./root/admin-openrc
```

2）列出可用的 meters

```
# ceilometer meter-list
```

如图 14-17 所示。

```
[root@controller ~]# ceilometer meter-list
+--------------------------------+-------+-----------+------------------------------------------------+---------+----------------------------------+
| Name                           | Type  | Unit      | Resource ID                                    | User ID | Project ID                       |
+--------------------------------+-------+-----------+------------------------------------------------+---------+----------------------------------+
| image                          | gauge | image     | bbac27a8-868e-4b85-89d8-d14df3e9233f            | None    | cec93d07f74a4346a3b2221241abb52e |
| image.size                     | gauge | B         | bbac27a8-868e-4b85-89d8-d14df3e9233f            | None    | cec93d07f74a4346a3b2221241abb52e |
| storage.containers.objects     | gauge | object    | cec93d07f74a4346a3b2221241abb52e/container1    | None    | cec93d07f74a4346a3b2221241abb52e |
| storage.containers.objects.size| gauge | B         | cec93d07f74a4346a3b2221241abb52e/container1    | None    | cec93d07f74a4346a3b2221241abb52e |
| storage.objects                | gauge | object    | 3ef3b5ed56084a8b9edfdf1a679a822f                | None    | 3ef3b5ed56084a8b9edfdf1a679a822f |
| storage.objects                | gauge | object    | 9ad59a65b5c44cc79146ef2c3b231c5e                | None    | 9ad59a65b5c44cc79146ef2c3b231c5e |
| storage.objects                | gauge | object    | cec93d07f74a4346a3b2221241abb52e                | None    | cec93d07f74a4346a3b2221241abb52e |
| storage.objects.containers     | gauge | container | 3ef3b5ed56084a8b9edfdf1a679a822f                | None    | 3ef3b5ed56084a8b9edfdf1a679a822f |
| storage.objects.containers     | gauge | container | 9ad59a65b5c44cc79146ef2c3b231c5e                | None    | 9ad59a65b5c44cc79146ef2c3b231c5e |
| storage.objects.containers     | gauge | container | cec93d07f74a4346a3b2221241abb52e                | None    | cec93d07f74a4346a3b2221241abb52e |
| storage.objects.outgoing.bytes | delta | B         | 3ef3b5ed56084a8b9edfdf1a679a822f                | None    | 3ef3b5ed56084a8b9edfdf1a679a822f |
| storage.objects.outgoing.bytes | delta | B         | 9ad59a65b5c44cc79146ef2c3b231c5e                | None    | 9ad59a65b5c44cc79146ef2c3b231c5e |
| storage.objects.outgoing.bytes | delta | B         | cec93d07f74a4346a3b2221241abb52e                | None    | cec93d07f74a4346a3b2221241abb52e |
| storage.objects.size           | gauge | B         | 3ef3b5ed56084a8b9edfdf1a679a822f                | None    | 3ef3b5ed56084a8b9edfdf1a679a822f |
| storage.objects.size           | gauge | B         | 9ad59a65b5c44cc79146ef2c3b231c5e                | None    | 9ad59a65b5c44cc79146ef2c3b231c5e |
| storage.objects.size           | gauge | B         | cec93d07f74a4346a3b2221241abb52e                | None    | cec93d07f74a4346a3b2221241abb52e |
+--------------------------------+-------+-----------+------------------------------------------------+---------+----------------------------------+
```

图 14-17 列出可用的 meters

3) 从镜像服务下载 CirrOS 镜像

\# IMAGE_ID = $ （glance image-list | grep 'cirros' | awk '{ print $ 2 } '）

\# glance image-download $ IMAGE_ID > /tmp/cirros.img

\# ll /tmp/cirros.img

如图 14-18 所示。

```
[root@controller ~]# IMAGE_ID=$(glance image-list | grep 'cirros' | awk '{ print $2 }')
[root@controller ~]#
[root@controller ~]# glance image-download $IMAGE_ID > /tmp/cirros.img
[root@controller ~]#
[root@controller ~]# ll /tmp/cirros.img
-rw-r--r-- 1 root root 13287936 3月  28 10:12 /tmp/cirros.img
[root@controller ~]#
```

图 14-18　从镜像服务下载 CirrOS 镜像

4) 再次列出可用的 meters 以验证镜像下载的检查

\# ceilometer meter-list

如图 14-19 所示。

```
[root@controller ~]# ceilometer meter-list
+----------------------------------+-------+-----------+----------------------------------------------+---------+
| Name                             | Type  | Unit      | Resource ID                                  | User ID |
| Project ID                       |       |           |                                              |         |
+----------------------------------+-------+-----------+----------------------------------------------+---------+
| image                            | gauge | image     | bbac27a8-868e-4b85-89d8-d14df3e9233f         | None    |
| cec93d07f74a4346a3b2221241abb52e |
| image.download                   | delta | B         | bbac27a8-868e-4b85-89d8-d14df3e9233f         | None    |
| cec93d07f74a4346a3b2221241abb52e |
| image.serve                      | delta | B         | bbac27a8-868e-4b85-89d8-d14df3e9233f         | None    |
| cec93d07f74a4346a3b2221241abb52e |
| image.size                       | gauge | B         | bbac27a8-868e-4b85-89d8-d14df3e9233f         | None    |
| cec93d07f74a4346a3b2221241abb52e |
| storage.containers.objects       | gauge | object    | cec93d07f74a4346a3b2221241abb52e/container1  | None    |
| cec93d07f74a4346a3b2221241abb52e |
| storage.containers.objects.size  | gauge | B         | cec93d07f74a4346a3b2221241abb52e/container1  | None    |
| cec93d07f74a4346a3b2221241abb52e |
| storage.objects                  | gauge | object    | 3ef3b5ed56084a8b9edfdf1a679a822f             | None    |
| 3ef3b5ed56084a8b9edfdf1a679a822f |
| storage.objects                  | gauge | object    | 9ad59a65b5c44cc79146ef2c3b231c5e             | None    |
| 9ad59a65b5c44cc79146ef2c3b231c5e |
| storage.objects                  | gauge | object    | cec93d07f74a4346a3b2221241abb52e             | None    |
| cec93d07f74a4346a3b2221241abb52e |
| storage.objects.containers       | gauge | container | 3ef3b5ed56084a8b9edfdf1a679a822f             | None    |
| 3ef3b5ed56084a8b9edfdf1a679a822f |
| storage.objects.containers       | gauge | container | 9ad59a65b5c44cc79146ef2c3b231c5e             | None    |
| 9ad59a65b5c44cc79146ef2c3b231c5e |
| storage.objects.containers       | gauge | container | cec93d07f74a4346a3b2221241abb52e             | None    |
| cec93d07f74a4346a3b2221241abb52e |
| storage.objects.outgoing.bytes   | delta | B         | 3ef3b5ed56084a8b9edfdf1a679a822f             | None    |
| 3ef3b5ed56084a8b9edfdf1a679a822f |
| storage.objects.outgoing.bytes   | delta | B         | 9ad59a65b5c44cc79146ef2c3b231c5e             | None    |
| 9ad59a65b5c44cc79146ef2c3b231c5e |
| storage.objects.outgoing.bytes   | delta | B         | cec93d07f74a4346a3b2221241abb52e             | None    |
| cec93d07f74a4346a3b2221241abb52e |
| storage.objects.size             | gauge | B         | 3ef3b5ed56084a8b9edfdf1a679a822f             | None    |
| 3ef3b5ed56084a8b9edfdf1a679a822f |
| storage.objects.size             | gauge | B         | 9ad59a65b5c44cc79146ef2c3b231c5e             | None    |
| 9ad59a65b5c44cc79146ef2c3b231c5e |
| storage.objects.size             | gauge | B         | cec93d07f74a4346a3b2221241abb52e             | None    |
| cec93d07f74a4346a3b2221241abb52e |
+----------------------------------+-------+-----------+----------------------------------------------+---------+
```

图 14-19　再次列出可用的 meters 以验证镜像下载的检查

5）从 image.download 表读取使用量统计值

ceilometer statistics-m image.download-p 60

如图 14-20 所示。

```
[root@controller ~]# ceilometer statistics -m image.download -p 60
+--------+--------------+----------------------------+----------------------------+------------+------------+------------+--------+
| Period | Period Start | Period End                 | Max        | Min        | Avg        | Sum    |
|        | Count | Duration | Duration Start        | Duration End               |            |            |            |        |
+--------+--------------+----------------------------+----------------------------+------------+------------+------------+--------+
| 60     | 2017-03-28T02:11:23.531000 | 2017-03-28T02:12:23.531000 | 13287936.0 | 13287936.0 | 13287936.0 | 1328
7936.0 | 1     | 0.0   | 2017-03-28T02:12:01.744000 | 2017-03-28T02:12:01.744000 |
+--------+--------------+----------------------------+----------------------------+------------+------------+------------+--------+
```

图 14-20 从 image.download 表读取使用量统计值

6）删除此前下载的 CirrOS 镜像文件

rm-rf /tmp/cirros.img

第 15 章　创建虚拟机实例

15.1　创建虚拟网络（控制节点）

在启动实例之前，必须创建必要的虚拟机网络设施。

本节主要讲述如何创建虚拟机实例所必须的虚拟网络。在控制节点上使用命令行（CLI）工具进行创建操作。

本书示例是基于私有网络（Self-service networks）的架构，但在创建私有网络之前，还是需要创建公共网络（Provider networks）的，它是私有网络的创建基础和前提。

公共网络（Provider networks）只需创建一个公共网络。而私有网络（Self-service networks）需同时创建一个公共网络和一个私有网络。

公共网络（Provider networks）包含一个使用公共虚拟网络（外部网络）的实例。这个网络通过 L2（桥/交换机）设备连接到物理网络设施，包括为实例提供 IP 地址的 DHCP 服务器。

私有网络（Self-service networks）包含一个使用公共虚拟网络的实例、一个使用私有虚拟网络（私有网络）的实例。在公共网络基础上，这个网络可以创建一个私有网络，并通过 NAT 连接到物理网络设施，包括一个 DHCP 服务器为实例分配 IP 地址。在私有网络上的实例可以自动连接到外部网络如互联网，但是，从外部网络如互联网访问实例需要配置浮动 IP。

15.1.1　配置网络

为简单起见，本书示例中使用控制节点的第 2 块网卡作为外部网络，以用于网络测试。

1）修改第 2 块网卡设备（eno33557248）的接口配置文件

```
# gedit /etc/sysconfig/network-scripts/ifcfg-eno33557248
```

修改或添加以下内容：

```
TYPE = Ethernet
BOOTPROTO = none
ONBOOT = yes
IPADDR = 203. 0. 113. 11
NETMASK = 255. 255. 255. 0
GATEWAY = 203. 0. 113. 11
```

2）重启网络服务

\# systemctl restart network. service

15.1.2 创建公共网络

公共网络（Provider networks）概览图，如图 15 – 1 所示。

Networking Option 1:Provider Networks Overview

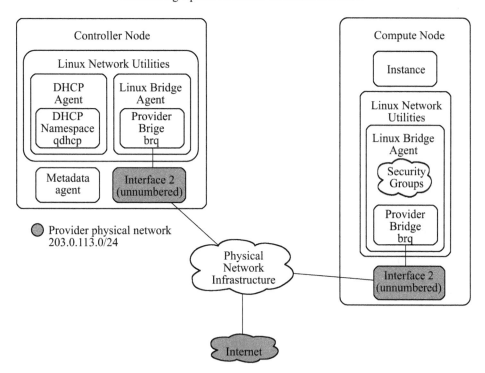

图 15 – 1 公共网络（Provider networks）概览图

公共网络（Provider networks）连接图，如图 15 – 2 所示。

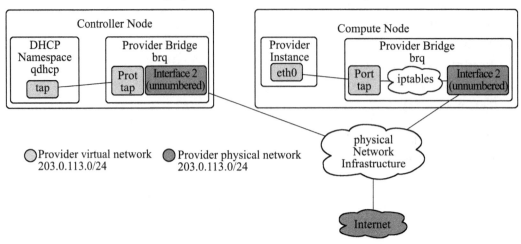

图 15 – 2 公共网络（Provider networks）连接图

必须由 admin 或者其他管理员权限用户创建这个网络，因为它直接连接到物理网络设施。

1）获得 admin 凭证来获取只有管理员才能执行的命令的访问权限

\# ./root/admin-openrc

2）创建网络

\# neutron net-create--shared--provider：physical_network provider \

　--provider：network_type flat provider

如图 15 – 3 所示。

```
[root@controller ~]# neutron net-create --shared --provider:physical_network provider \
> --provider:network_type flat provider
Created a new network:
+---------------------------+--------------------------------------+
| Field                     | Value                                |
+---------------------------+--------------------------------------+
| admin_state_up            | True                                 |
| availability_zone_hints   |                                      |
| availability_zones        |                                      |
| created_at                | 2017-04-05T07:27:32                  |
| description               |                                      |
| id                        | e57af6ca-6f43-4aba-8d74-bc9ef537eea9 |
| ipv4_address_scope        |                                      |
| ipv6_address_scope        |                                      |
| mtu                       | 1500                                 |
| name                      | provider                             |
| port_security_enabled     | True                                 |
| provider:network_type     | flat                                 |
| provider:physical_network | provider                             |
| provider:segmentation_id  |                                      |
| router:external           | False                                |
| shared                    | True                                 |
| status                    | ACTIVE                               |
| subnets                   |                                      |
| tags                      |                                      |
| tenant_id                 | cec93d07f74a4346a3b2221241abb52e     |
| updated_at                | 2017-04-05T07:27:32                  |
+---------------------------+--------------------------------------+
```

图 15 – 3 创建网络

选项说明：

--shared 选项：允许所有项目（租户）使用该虚拟网络。

--provider：physical_network provider 和--provider：network_type 选项：flat 类型的虚拟网络连接到 provider 类型的物理网络的网络接口（第 2 块网卡为 eno33557248）。

相关网络配置信息在以下配置文件中均已定义。

在 Modular Layer 2（ML2）插件配置文件 /etc/neutron/plugins/ml2/ml2_conf.ini 中的 [ml2_type_flat] 部分，定义了公共虚拟网络为 flat 网络：

[ml2_type_flat]

...

flat_networks = provider

在 Linuxbridge 代理配置文件 /etc/neutron/plugins/ml2/linuxbridge_agent.ini 中的 [linux_bridge] 部分，定义了将公共虚拟网络和公共物理网络接口对应起来：

[linux_bridge]

...

physical_interface_mappings = provider：eno33557248

3）在网络上创建一个子网

命令格式：

neutron subnet-create--name provider \

  --allocation-pool start = START_IP_ADDRESS，end = END_IP_ADDRESS \

  --dns-nameserver DNS_RESOLVER \

  --gateway PROVIDER_NETWORK_GATEWAY \

  provider PROVIDER_NETWORK_CIDR

选项说明：

使用物理网络的子网 CIDR 标记替换 PROVIDER_NETWORK_CIDR。

将 START_IP_ADDRESS 和 END_IP_ADDRESS 使用想分配给实例的子网网段的第一个和最后一个 IP 地址。这个范围不能包括任何已经使用的 IP 地址。

将 DNS_RESOLVER 替换为 DNS 解析服务的 IP 地址。

将 PUBLIC_NETWORK_GATEWAY 替换为公共网络的网关。

命令示例：

公共网络 203.0.113.0/24 的网关为 203.0.113.1，DHCP 服务为每个实例分配 IP，IP

地址范围从 203.0.113.101 到 203.0.113.250，所有实例的 DNS 使用 8.8.4.4。

neutron subnet-create--name provider \

    --allocation-pool start = 203.0.113.101，end = 203.0.113.250 \

    --dns-nameserver 8.8.4.4 \

    --gateway 203.0.113.1 \

    provider 203.0.113.0/24

如图 15 -4 所示。

```
[root@controller ~]# neutron subnet-create --name provider \
> --allocation-pool start=203.0.113.101,end=203.0.113.250 \
> --dns-nameserver 8.8.4.4 \
> --gateway 203.0.113.1 \
> provider 203.0.113.0/24
Created a new subnet:
+--------------------+------------------------------------------------+
| Field              | Value                                          |
+--------------------+------------------------------------------------+
| allocation_pools   | {"start": "203.0.113.101", "end": "203.0.113.250"} |
| cidr               | 203.0.113.0/24                                 |
| created_at         | 2017-04-05T07:29:38                            |
| description        |                                                |
| dns_nameservers    | 8.8.4.4                                        |
| enable_dhcp        | True                                           |
| gateway_ip         | 203.0.113.1                                    |
| host_routes        |                                                |
| id                 | e2c834a7-c2f5-4b5c-abae-4bd0f0177604            |
| ip_version         | 4                                              |
| ipv6_address_mode  |                                                |
| ipv6_ra_mode       |                                                |
| name               | provider                                       |
| network_id         | e57af6ca-6f43-4aba-8d74-bc9ef537eea9            |
| subnetpool_id      |                                                |
| tenant_id          | cec93d07f74a4346a3b2221241abb52e               |
| updated_at         | 2017-04-05T07:29:38                            |
+--------------------+------------------------------------------------+
```

图 15 -4　配置 IP 和 DNS 示例

15.1.3　创建私有网络

1. 私有网络（Self-service networks）概述，如图 15 -5 所示。

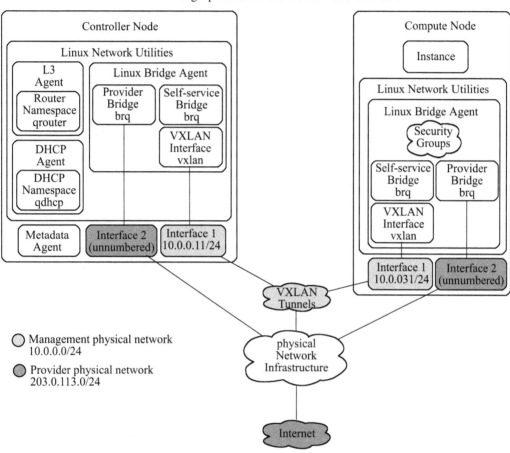

图 15-5 私有网络（Self-service networks）概述

2. 私有网络（Self-service networks）连接，如图 15-6 所示

在创建私有网络之前，必须创建公共网络 provider。

demo 或者其他非管理员用户也可以创建这个网络，因为它只在 demo 项目中提供对实例的访问。

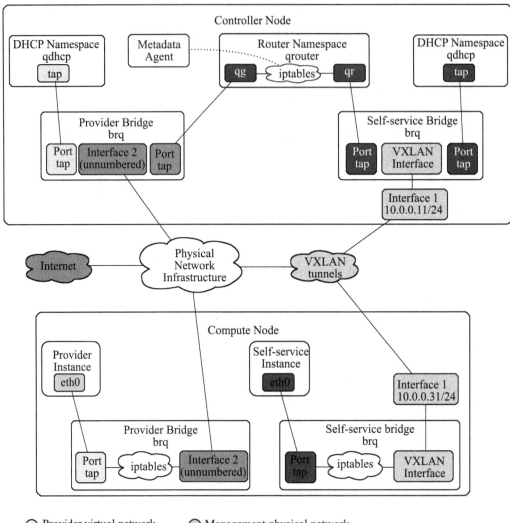

图 15-6 私有网络（Self-service networks）连接

1）获得 demo 凭证来获取只有非管理员才能执行的命令的访问权限
./root/demo-openrc
2）创建名为 selfservice 的私有网络
neutron net-create selfservice

如图 15-7 所示。

```
[root@controller ~]# neutron net-create selfservice
Created a new network:
+---------------------------+--------------------------------------+
| Field                     | Value                                |
+---------------------------+--------------------------------------+
| admin_state_up            | True                                 |
| availability_zone_hints   |                                      |
| availability_zones        |                                      |
| created_at                | 2017-04-05T07:31:35                  |
| description               |                                      |
| id                        | c97a05d9-b771-49e7-884e-559270aba95b |
| ipv4_address_scope        |                                      |
| ipv6_address_scope        |                                      |
| mtu                       | 1450                                 |
| name                      | selfservice                          |
| port_security_enabled     | True                                 |
| router:external           | False                                |
| shared                    | False                                |
| status                    | ACTIVE                               |
| subnets                   |                                      |
| tags                      |                                      |
| tenant_id                 | 9ad59a65b5c44cc79146ef2c3b231c5e     |
| updated_at                | 2017-04-05T07:31:35                  |
+---------------------------+--------------------------------------+
```

图 15-7 创建名为 self-service 的私有网络

非特权用户一般不能在这个命令制定更多参数。服务会自动从 ml2 配置文件信息中选择参数：

在 Modular Layer 2（ML2）插件配置文件 /etc/neutron/plugins/ml2/ml2_conf.ini 中的 [ml2] 和 [ml2_type_flat] 部分，定义了私有网络的类型和范围：

[ml2]
tenant_network_types = vxlan

[ml2_type_vxlan]
vni_ranges = 1：1000

3）在网络上创建一个子网
命令格式：
neutron subnet-create--name selfservice \
 --dns-nameserver DNS_RESOLVER \
 --gateway SELFSERVICE_NETWORK_GATEWAY \
 selfservice SELFSERVICE_NETWORK_CIDR

将 DNS_RESOLVER 替换为 DNS 解析服务的 IP 地址，如图 15-8 所示。
将 PRIVATE_NETWORK_GATEWAY 替换为私有网络的网关。
将 PRIVATE_NETWORK_CIDR 替换为私有网络的子网。

命令示例：

私有网络 172.16.1.0/24 使用网关 172.16.1.1。DHCP 服务负责为每个实例从 172.16.1.2 到 172.16.1.254 中分配 IP 地址。所有实例的 DNS 使用 172.16.1.1。

```
# neutron subnet-create --name selfservice \
    --dns-nameserver 172.16.1.1 \
    --gateway 172.16.1.1 \
    selfservice 172.16.1.0/24
```

如图 15-8 所示。

```
[root@controller ~]# neutron subnet-create --name selfservice \
> --dns-nameserver 172.16.1.1 \
> --gateway 172.16.1.1 \
> selfservice 172.16.1.0/24
Created a new subnet:
+-------------------+------------------------------------------------+
| Field             | Value                                          |
+-------------------+------------------------------------------------+
| allocation_pools  | {"start": "172.16.1.2", "end": "172.16.1.254"} |
| cidr              | 172.16.1.0/24                                  |
| created_at        | 2017-04-05T07:32:05                            |
| description       |                                                |
| dns_nameservers   | 172.16.1.1                                     |
| enable_dhcp       | True                                           |
| gateway_ip        | 172.16.1.1                                     |
| host_routes       |                                                |
| id                | cb4abfb4-dfc8-4725-8b8d-807bd536cb4f           |
| ip_version        | 4                                              |
| ipv6_address_mode |                                                |
| ipv6_ra_mode      |                                                |
| name              | selfservice                                    |
| network_id        | c97a05d9-b771-49e7-884e-559270aba95b           |
| subnetpool_id     |                                                |
| tenant_id         | 9ad59a65b5c44cc79146ef2c3b231c5e               |
| updated_at        | 2017-04-05T07:32:05                            |
+-------------------+------------------------------------------------+
```

图 15-8　配置 selfservice 私有网络示例

15.1.3.1　创建路由

私有网络通过虚拟路由来连接到公共网络，以双向 NAT 最为典型。每个路由包含至少一个连接到私有网络的接口以及一个连接到公共网络的网关的接口。

公共网络必须包括 router：external 选项，以使路由连接到外部网络，比如互联网。admin 或者其他权限用户在网络创建时必须包括这个选项，也可以之后再添加。

本书示例中环境把公共网络设置成 router：external。

1）获得 admin 凭证来获取只有管理员才能执行的命令的访问权限

```
# ./root/admin-openrc
```

2）添加 router：external 到 provider 网络上

```
# neutron net-update provider --router：external
```

如图 15-9 所示。

neutron router-create router

```
[root@controller ~]# neutron net-update provider --router:external
Updated network: provider
```

图 15-9　添加 router：external 到 provider 网络上

3) 获得 demo 凭证来获取只有非管理员才能执行的命令的访问权限

./root/demo-openrc

4) 创建路由

如图 15-10 所示。

```
[root@controller ~]# neutron router-create router
Created a new router:
+-------------------------+--------------------------------------+
| Field                   | Value                                |
+-------------------------+--------------------------------------+
| admin_state_up          | True                                 |
| availability_zone_hints |                                      |
| availability_zones      |                                      |
| description             |                                      |
| external_gateway_info   |                                      |
| id                      | f4557326-2e0e-49b4-b8b9-ce4796010cc9 |
| name                    | router                               |
| routes                  |                                      |
| status                  | ACTIVE                               |
| tenant_id               | 9ad59a65b5c44cc79146ef2c3b231c5e     |
+-------------------------+--------------------------------------+
```

图 15-10　创建路由

5) 给路由器添加一个私网子网的接口

neutron router-interface-add router selfservice

如图 15-11 所示。

```
[root@controller ~]# neutron router-interface-add router selfservice
Added interface e808cbf4-59c2-4f7c-b7ac-790532bac8d6 to router router.
```

图 15-11　给路由器添加一个私网子网的接口

6) 给路由器设置公共网络的网关

neutron router-gateway-set router provider

如图 15-12 所示。

```
[root@controller ~]# neutron router-gateway-set router provider
Set gateway for router router
```

图 15-12　给路由器设置公共网络的网关

15.1.3.2 验证操作

1) 获得 admin 凭证来获取只有管理员才能执行的命令的访问权限

./root/admin-openrc

2) 列出网络命名空间

ip netns

如图 15-13 所示。

```
[root@controller ~]# ip netns
qrouter-f4557326-2e0e-49b4-b8b9-ce4796010cc9 (id: 2)
qdhcp-c97a05d9-b771-49e7-884e-559270aba95b (id: 1)
qdhcp-e57af6ca-6f43-4aba-8d74-bc9ef537eea9 (id: 0)
```

图 15-13 列出网络命名空间

本书示例中环境可以看到一个 qrouter 命名空间和两个 qdhcp 命名空间：

3) 列出网络路由器

neutron router-list

如图 15-14 所示。

```
[root@controller ~]# neutron router-list
```

id	name	external_gateway_info	distributed	ha
f4557326-2e0e-49b4-b8b9-ce4796010cc9	router	{"network_id": "e57af6ca-6f43-4aba-8d74-bc9ef537eea9", "enable_snat": true, "external_fixed_ips": [{"subnet_id": "e2c834a7-c2f5-4b5c-abae-4bd0f0177604", "ip_address": "203.0.113.102"}]}	False	False

图 15-14 列出网络路由器

4) 列出路由器上的端口来确定公网网关的 IP 地址

neutron router-port-list router

如图 15-15 所示。

```
[root@controller ~]# neutron router-port-list router
```

id	name	mac_address	fixed_ips
d0d2a622-33e9-44e4-8b71-f53d9da70869		fa:16:3e:d2:5a:43	{"subnet_id": "e2c834a7-c2f5-4b5c-abae-4bd0f0177604", "ip_address": "203.0.113.102"}
e808cbf4-59c2-4f7c-b7ac-790532bac8d6		fa:16:3e:76:f1:72	{"subnet_id": "cb4abfb4-dfc8-4725-8b8d-807bd536cb4f", "ip_address": "172.16.1.1"}

图 15-15 列出路由器上的端口来确定公网网关的 IP 地址

15.2 创建 m1.nano 规格主机（控制节点）

默认最小规格的主机需要 512MB 内存，对于环境中计算节点内存不足 4GB 的，推荐创建只需要 64MB 的 m1.nano 规格的主机。若单纯为了测试的目的，则请使用 m1.nano 规格的主机来加载 CirrOS 镜像。

1）获得 admin 凭证来获取只有管理员才能执行的命令的访问权限

./root/admin-openrc

2）列出可用主机类型

openstack flavor list

如图 15-16 所示。

```
[root@controller ~]# openstack flavor list
+----+-----------+-------+------+-----------+-------+-----------+
| ID | Name      | RAM   | Disk | Ephemeral | VCPUs | Is Public |
+----+-----------+-------+------+-----------+-------+-----------+
| 1  | m1.tiny   | 512   | 1    | 0         | 1     | True      |
| 2  | m1.small  | 2048  | 20   | 0         | 1     | True      |
| 3  | m1.medium | 4096  | 40   | 0         | 2     | True      |
| 4  | m1.large  | 8192  | 80   | 0         | 4     | True      |
| 5  | m1.xlarge | 16384 | 160  | 0         | 8     | True      |
+----+-----------+-------+------+-----------+-------+-----------+
```

图 15-16 列出可用主机类型

3）创建只需要 64MB 的 m1.nano 规格的主机

openstack flavor create--id 0--vcpus 1--ram 64--disk 1 m1.nano

如图 15-17 所示。

```
[root@controller ~]# openstack flavor create --id 0 --vcpus 1 --ram 64 --disk 1 m1.nano
+----------------------------+---------+
| Field                      | Value   |
+----------------------------+---------+
| OS-FLV-DISABLED: disabled  | False   |
| OS-FLV-EXT-DATA: ephemeral | 0       |
| disk                       | 1       |
| id                         | 0       |
| name                       | m1.nano |
| os-flavor-access: is_public| True    |
| ram                        | 64      |
| rxtx_factor                | 1.0     |
| swap                       |         |
| vcpus                      | 1       |
+----------------------------+---------+
```

图 15-17 创建只需要 64MB 的 m1.nano 规格的主机

4）列出可用主机类型

openstack flavor list

如图 15-18 所示。

```
[root@controller ~]# openstack flavor list
+----+-----------+-------+------+-----------+-------+-----------+
| ID | Name      | RAM   | Disk | Ephemeral | VCPUs | Is Public |
+----+-----------+-------+------+-----------+-------+-----------+
| 0  | m1.nano   | 64    | 1    | 0         | 1     | True      |
| 1  | m1.tiny   | 512   | 1    | 0         | 1     | True      |
| 2  | m1.small  | 2048  | 20   | 0         | 1     | True      |
| 3  | m1.medium | 4096  | 40   | 0         | 2     | True      |
| 4  | m1.large  | 8192  | 80   | 0         | 4     | True      |
| 5  | m1.xlarge | 16384 | 160  | 0         | 8     | True      |
+----+-----------+-------+------+-----------+-------+-----------+
```

图 15 – 18　列出可用主机类型

15.3　生成密钥对（控制节点）

大部分云镜像支持公共密钥认证而不是传统的密码认证。在启动实例前，必须添加一个公共密钥到计算服务。

1）获得 demo 凭证来获取只有非管理员才能执行的命令的访问权限

＃．/root/demo-openrc

2）生成和添加密钥对

echo | ssh-keygen-q-N " "

openstack keypair create--public-key ~/.ssh/id_rsa.pub mykey

如图 15 – 19 所示。

```
[root@controller ~]# openstack keypair create --public-key ~/.ssh/id_rsa.pub mykey
+-------------+-------------------------------------------------+
| Field       | Value                                           |
+-------------+-------------------------------------------------+
| fingerprint | 3b:fb:21:e8:69:f4:11:da:96:c6:85:e1:57:2f:d1:1e |
| name        | mykey                                           |
| user_id     | 455de29c0a3a4b86ac22b9f3f0b8ec67                |
+-------------+-------------------------------------------------+
```

图 15 – 19　生成和添加密钥对

3）验证公钥的添加

openstack keypair list

如图 15 – 20 所示。

```
[root@controller ~]# openstack keypair list
+-------+-------------------------------------------------+
| Name  | Fingerprint                                     |
+-------+-------------------------------------------------+
| mykey | 3b:fb:21:e8:69:f4:11:da:96:c6:85:e1:57:2f:d1:1e |
+-------+-------------------------------------------------+
```

图 15 – 20　验证公钥的添加

15.4 为 default 安全组添加规则（控制节点）

默认情况下，OpenStack 环境中包含一个 default 安全组，它适用于所有实例并且包括拒绝远程访问实例的防火墙规则。对诸如 CirrOS 这样的 Linux 镜像，推荐至少允许 ICMP（ping）和安全 shell（SSH）规则。

1) 获得 demo 凭证来获取只有非管理员才能执行的命令的访问权限

    # ./root/demo-openrc

2) 为 default 安全组添加允许 ICMP（ping）规则

    # openstack security group rule create--proto icmp default

如图 15-21 所示。

```
[root@controller ~]# openstack security group rule create --proto icmp default
+-------------------+--------------------------------------+
| Field             | Value                                |
+-------------------+--------------------------------------+
| id                | 37b7f0bd-46b8-4420-b304-54215800872d |
| ip_protocol       | icmp                                 |
| ip_range          | 0.0.0.0/0                            |
| parent_group_id   | f34fbc53-79fe-4ef7-b369-3b80eb46918f |
| port_range        |                                      |
| remote_security_group |                                  |
+-------------------+--------------------------------------+
```

图 15-21 为 default 安全组添加允许 ICMP（ping）规则

3) 为 default 安全组添加允许安全 shell（SSH）规则

    # openstack security group rule create--proto tcp--dst-port 22 default

如图 15-22 所示。

```
[root@controller ~]# openstack security group rule create --proto tcp --dst-port 22 default
+-------------------+--------------------------------------+
| Field             | Value                                |
+-------------------+--------------------------------------+
| id                | e64e2be1-380c-41f0-97ff-872df4736356 |
| ip_protocol       | tcp                                  |
| ip_range          | 0.0.0.0/0                            |
| parent_group_id   | f34fbc53-79fe-4ef7-b369-3b80eb46918f |
| port_range        | 22:22                                |
| remote_security_group |                                  |
+-------------------+--------------------------------------+
```

图 15-22 为 default 安全组添加允许安全 shell（SSH）规则

15.5 创建虚拟机实例（控制节点）

如果仅创建了公共网络，只能在公共网络上创建实例。如果创建了私有网络，可以在公共网络或私有网络上创建实例。

15.5.1 在公共网络上创建虚拟机实例

启动一台实例，必须至少指定一个类型、镜像名称、网络、安全组、密钥和实例名称。

15.5.1.1 确定实例选项

1）获得 demo 凭证来获取只有非管理员才能执行的命令的访问权限：

\# ./root/demo-openrc

2）列出可用主机类型

\# openstack flavor list

如图 15-23 所示。

```
[root@controller ~]# openstack flavor list
+----+-----------+-------+------+-----------+-------+-----------+
| ID | Name      | RAM   | Disk | Ephemeral | VCPUs | Is Public |
+----+-----------+-------+------+-----------+-------+-----------+
| 0  | m1.nano   |   64  |   1  |     0     |   1   |   True    |
| 1  | m1.tiny   |  512  |   1  |     0     |   1   |   True    |
| 2  | m1.small  |  2048 |  20  |     0     |   1   |   True    |
| 3  | m1.medium |  4096 |  40  |     0     |   2   |   True    |
| 4  | m1.large  |  8192 |  80  |     0     |   4   |   True    |
| 5  | m1.xlarge | 16384 | 160  |     0     |   8   |   True    |
+----+-----------+-------+------+-----------+-------+-----------+
```

图 15-23　列出可用主机类型

一个实例需指定虚拟机资源的大致分配，包括处理器、内存和存储。

本书中示例中虚拟机实例将使用 m1.nano 类型的主机。

3）列出可用镜像

\# openstack image list

如图 15-24 所示。

图 15-24　列出可用镜像

本书中示例中虚拟机实例将使用 cirros 镜像。

4）列出可用网络

\# openstack network list

如图 15-25 所示。

```
[root@controller ~]# openstack network list
+--------------------------------------+-------------+--------------------------------------+
| ID                                   | Name        | Subnets                              |
+--------------------------------------+-------------+--------------------------------------+
| e57af6ca-6f43-4aba-8d74-bc9ef537eea9 | provider    | e2c834a7-c2f5-4b5c-abae-4bd0f0177604 |
| c97a05d9-b771-49e7-884e-559270aba95b | selfservice | cb4abfb4-dfc8-4725-8b8d-807bd536cb4f |
+--------------------------------------+-------------+--------------------------------------+
```

图 15-25 列出可用网络

在使用网络时，必须使用网络 ID，而不是网络的名称。

5）列出可用网络安全组

\# openstack security group list

如图 15-26 所示。

```
[root@controller ~]# openstack security group list
+--------------------------------------+---------+------------------------+----------------------------------+
| ID                                   | Name    | Description            | Project                          |
+--------------------------------------+---------+------------------------+----------------------------------+
| f34fbc53-79fe-                       | default | Default security group | 9ad59a65b5c44cc79146ef2c3b2      |
| 4ef7-b369-3b80eb46918f               |         |                        | 31c5e                            |
+--------------------------------------+---------+------------------------+----------------------------------+
```

图 15-26 列出可用网络安全组

本书中示例中虚拟机实例将使用 default 安全组。

15.5.1.2 创建虚拟机实例

1）创建虚拟机实例

\# openstack server create \\

　　--flavor m1.nano \\

　　--image cirros \\

　　--nic net-id=PROVIDER_NET_ID \\

　　--security-group default \\

　　--key-name mykey provider-instance

使用公共网络 ID 替换 PROVIDER_NET_ID。

本书中示例中虚拟机实例将使用 default 安全组，如图 15-27 所示。

```
[root@controller ~]# openstack server create \
> --flavor m1.nano \
> --image cirros \
> --nic net-id=e57af6ca-6f43-4aba-8d74-bc9ef537eea9 \
> --security-group default \
> --key-name mykey provider-instance
+-------------------------------------+-----------------------------------------------+
| Field                               | Value                                         |
+-------------------------------------+-----------------------------------------------+
| OS-DCF: diskConfig                  | MANUAL                                        |
| OS-EXT-AZ: availability_zone        |                                               |
| OS-EXT-STS: power_state             | 0                                             |
| OS-EXT-STS: task_state              | scheduling                                    |
| OS-EXT-STS: vm_state                | building                                      |
| OS-SRV-USG: launched_at             | None                                          |
| OS-SRV-USG: terminated_at           | None                                          |
| accessIPv4                          |                                               |
| accessIPv6                          |                                               |
| addresses                           |                                               |
| adminPass                           | zeYMrtWB76kY                                  |
| config_drive                        |                                               |
| created                             | 2017-04-05T07:44:22Z                          |
| flavor                              | m1.nano (0)                                   |
| hostId                              |                                               |
| id                                  | 0ab30d71-c566-4928-bb8c-9c1fdd0eba15          |
| image                               | cirros (bbac27a8-868e-4b85-89d8-d14df3e9233f) |
| key_name                            | mykey                                         |
| name                                | provider-instance                             |
| os-extended-volumes: volumes_attached | []                                          |
| progress                            | 0                                             |
| project_id                          | 9ad59a65b5c44cc79146ef2c3b231c5e              |
| properties                          |                                               |
| security_groups                     | [{u'name': u'default'}]                       |
| status                              | BUILD                                         |
| updated                             | 2017-04-05T07:44:22Z                          |
| user_id                             | 455de29c0a3a4b86ac22b9f3f0b8ec67              |
+-------------------------------------+-----------------------------------------------+
```

图 15-27　创建虚拟机实例

如果在 OpenStack 环境中仅有一个公共网络，那么此时可以省去-nic 选项，因为 OpenStack 会自动选择这个唯一可用的网络。

2）检查虚拟机实例的状态

openstack server list

如图 15-28 所示。

```
[root@controller ~]# openstack server list
+--------------------------------------+-------------------+--------+-------------------------+
| ID                                   | Name              | Status | Networks                |
+--------------------------------------+-------------------+--------+-------------------------+
| 0ab30d71-c566-4928-bb8c-9c1fdd0eba15 | provider-instance | ACTIVE | provider=203.0.113.103  |
+--------------------------------------+-------------------+--------+-------------------------+
```

图 15-28　检查虚拟机实例的状态

当虚拟机实例构建过程完全成功后，状态会从 BUILD 变为 ACTIVE。

15.5.1.3 使用虚拟控制台访问实例

1）获取虚拟机实例的 VNC 会话 URL

\# openstack console url show provider-instance

如图 15-29 所示。

```
[root@controller ~]# openstack console url show provider-instance
+-------+--------------------------------------------------------------------------------+
| Field | Value                                                                          |
+-------+--------------------------------------------------------------------------------+
| type  | novnc                                                                          |
| url   | http://controller:6080/vnc_auto.html?token=bdedfa23-d4c0-4155-99d5-fdff2c6f4913 |
+-------+--------------------------------------------------------------------------------+
```

图 15-29 获取虚拟机实例的 VNC 会话 URL

2）从 Web 浏览器访问 VNC 会话 URL

如图 15-30 所示。

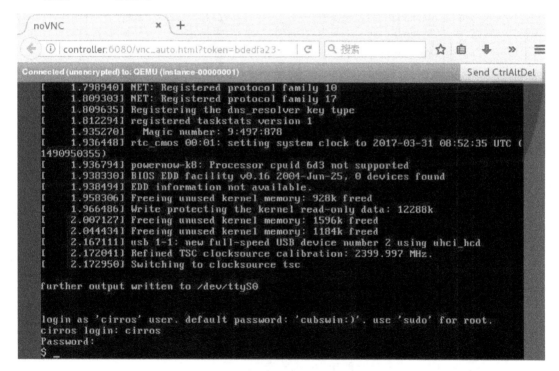

图 15-30 从 Web 浏览器访问 VNC 会话 URL

CirrOS 镜像的默认用户名为 cirros，默认密码为 cubswin。

15.5.2 在私有网络上创建虚拟机实例

启动一台实例,必须至少指定一个类型、镜像名称、网络、安全组、密钥和实例名称。

15.5.2.1 确定实例选项

1) 获得 demo 凭证来获取只有非管理员才能执行的命令的访问权限

. /root/demo-openrc

2) 列出可用主机类型

openstack flavor list

如图 15-31 所示。

```
[root@controller ~]# openstack flavor list
+----+-----------+-------+------+-----------+-------+-----------+
| ID | Name      | RAM   | Disk | Ephemeral | VCPUs | Is Public |
+----+-----------+-------+------+-----------+-------+-----------+
| 0  | m1.nano   | 64    | 1    | 0         | 1     | True      |
| 1  | m1.tiny   | 512   | 1    | 0         | 1     | True      |
| 2  | m1.small  | 2048  | 20   | 0         | 1     | True      |
| 3  | m1.medium | 4096  | 40   | 0         | 2     | True      |
| 4  | m1.large  | 8192  | 80   | 0         | 4     | True      |
| 5  | m1.xlarge | 16384 | 160  | 0         | 8     | True      |
+----+-----------+-------+------+-----------+-------+-----------+
```

图 15-31 列出可用主机类型

一个实例需指定虚拟机资源的大致分配,包括处理器、内存和存储。

本书中示例中虚拟机实例将使用 m1.nano 类型的主机。

3) 列出可用镜像

openstack image list

如图 15-32 所示。

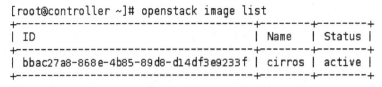

图 15-32 列出可用镜像

本书中示例中虚拟机实例将使用 cirros 镜像。

4) 列出可用网络

openstack network list

如图 15-33 所示。

```
[root@controller ~]# openstack network list
+--------------------------------------+-------------+--------------------------------------+
| ID                                   | Name        | Subnets                              |
+--------------------------------------+-------------+--------------------------------------+
| e57af6ca-6f43-4aba-8d74-bc9ef537eea9 | provider    | e2c834a7-c2f5-4b5c-abae-4bd0f0177604 |
| c97a05d9-b771-49e7-884e-559270aba95b | selfservice | cb4abfb4-dfc8-4725-8b8d-807bd536cb4f |
+--------------------------------------+-------------+--------------------------------------+
```

图 15-33 列出可用网络

在使用网络时，必须使用网络 ID，而不是网络的名称。

5）列出可用网络安全组

openstack security group list

如图 15-34 所示。

```
[root@controller ~]# openstack security group list
+--------------------------------------+---------+------------------------+--------------------------------+
| ID                                   | Name    | Description            | Project                        |
+--------------------------------------+---------+------------------------+--------------------------------+
| f34fbc53-79fe-                       | default | Default security group | 9ad59a65b5c44cc79146ef2c3b2    |
| 4ef7-b369-3b80eb46918f               |         |                        | 31c5e                          |
+--------------------------------------+---------+------------------------+--------------------------------+
```

图 15-34 列出可用网络安全组

本书中示例中虚拟机实例将使用 default 安全组。

15.5.2.2 创建虚拟机实例

1）获得 demo 凭证来获取只有非管理员才能执行的命令的访问权限

./root/demo-openrc

2）创建虚拟机实例

openstack server create \

 --flavor m1.nano \

 --image cirros \

 --nic net-id=SELFSERVICE_NET_ID \

 --security-group default \

 --key-name mykey selfservice-instance

请使用私有网络 ID 替换 SELFSERVICE_NET_ID。

本书中示例虚拟机实例将使用 default 安全组。

如图 15-35 所示。

```
[root@controller ~]# openstack server create \
> --flavor m1.nano \
> --image cirros \
> --nic net-id=c97a05d9-b771-49e7-884e-559270aba95b \
> --security-group default \
> --key-name mykey selfservice-instance
+--------------------------------------+-----------------------------------------------+
| Field                                | Value                                         |
+--------------------------------------+-----------------------------------------------+
| OS-DCF: diskConfig                   | MANUAL                                        |
| OS-EXT-AZ: availability_zone         |                                               |
| OS-EXT-STS: power_state              | 0                                             |
| OS-EXT-STS: task_state               | None                                          |
| OS-EXT-STS: vm_state                 | building                                      |
| OS-SRV-USG: launched_at              | None                                          |
| OS-SRV-USG: terminated_at            | None                                          |
| accessIPv4                           |                                               |
| accessIPv6                           |                                               |
| addresses                            |                                               |
| adminPass                            | Y8AgwNtEeTg8                                  |
| config_drive                         |                                               |
| created                              | 2017-04-05T07:46:53Z                          |
| flavor                               | m1.nano (0)                                   |
| hostId                               |                                               |
| id                                   | fb9841eb-daf5-4b5b-be74-3cc051b05c67          |
| image                                | cirros (bbac27a8-868e-4b85-89d8-d14df3e9233f) |
| key_name                             | mykey                                         |
| name                                 | selfservice-instance                          |
| os-extended-volumes: volumes_attached| []                                            |
| progress                             | 0                                             |
| project_id                           | 9ad59a65b5c44cc79146ef2c3b231c5e              |
| properties                           |                                               |
| security_groups                      | [{u'name': u'default'}]                       |
| status                               | BUILD                                         |
| updated                              | 2017-04-05T07:46:54Z                          |
| user_id                              | 455de29c0a3a4b86ac22b9f3f0b8ec67              |
+--------------------------------------+-----------------------------------------------+
```

图 15-35 创建虚拟机实例

3）检查虚拟机实例的状态

\# openstack server list

如图 15-36 所示。

```
[root@controller ~]# openstack server list
+--------------------------------------+----------------------+--------+-------------------------+
| ID                                   | Name                 | Status | Networks                |
+--------------------------------------+----------------------+--------+-------------------------+
| fb9841eb-daf5-4b5b-be74-3cc051b05c67 | selfservice-instance | ACTIVE | selfservice=172.16.1.3  |
| 0ab30d71-c566-4928-bb8c-9c1fdd0eba15 | provider-instance    | ACTIVE | provider=203.0.113.103  |
+--------------------------------------+----------------------+--------+-------------------------+
```

图 15-36 检查虚拟机实例的状态

当虚拟机实例构建过程完全成功后，状态会从 BUILD 变为 ACTIVE。

15.5.2.3 使用虚拟控制台访问实例

1）获得 demo 凭证来获取只有非管理员才能执行的命令的访问权限

\# ./root/demo-openrc

2）获取虚拟机实例的 VNC 会话 URL

\# openstack console url show selfservice-instance

如图 15-37 所示。

```
[root@controller ~]# openstack console url show selfservice-instance
+-------+-------------------------------------------------------------------------+
| Field | Value                                                                   |
+-------+-------------------------------------------------------------------------+
| type  | novnc                                                                   |
| url   | http://controller:6080/vnc_auto.html?token=805e63b6-ce0a-47a7-b81d-3f6e9b9de6b4 |
+-------+-------------------------------------------------------------------------+
```

图 15-37　获取虚拟机实例的 VNC 会话 URL

3) 从 Web 浏览器访问 VNC 会话 URL

如图 15-38 所示。

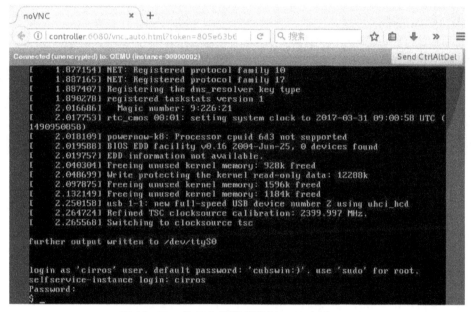

图 15-38　从 Web 浏览器访问 VNC 会话 URL

CirrOS 镜像的默认用户名为 cirros，默认密码为 cubswin:)。

4) 在虚拟机实例控制台中，验证实例是否从 DHCP 中获取 IP 地址

$ ifconfig

如图 15-39 所示。

图 15-39　验证实例是否从 DHCP 中获取 IP 地址

5）在虚拟机实例控制台中，验证实例的路由信息

$ sudo ip route

如图 15 -40 所示。

```
$ sudo ip route
default via 172.16.1.1 dev eth0
169.254.169.254 via 172.16.1.1 dev eth0
172.16.1.0/24 dev eth0  src 172.16.1.3
```

图 15 -40　验证实例的路由信息

6）在虚拟机实例控制台中，验证能否 ping 通私有网络的网关

$ ping-c 4 172.16.1.1

如图 15 -41 所示。

```
$ ping -c 4 172.16.1.1
PING 172.16.1.1 (172.16.1.1): 56 data bytes
64 bytes from 172.16.1.1: seq=0 ttl=64 time=12.738 ms
64 bytes from 172.16.1.1: seq=1 ttl=64 time=1.790 ms
64 bytes from 172.16.1.1: seq=2 ttl=64 time=0.806 ms
64 bytes from 172.16.1.1: seq=3 ttl=64 time=1.182 ms

--- 172.16.1.1 ping statistics ---
4 packets transmitted, 4 packets received, 0% packet loss
round-trip min/avg/max = 0.806/4.129/12.738 ms
```

图 15 -41　验证能否 ping 通私有网络网关

15.5.2.4　验证能否远程访问实例

1）获得 demo 凭证来获取只有非管理员才能执行的命令的访问权限

#. /root/demo-openrc

2）在私有网络上创建浮动 IP 地址池

openstack ip floating create provider

如图 15 -42 所示。

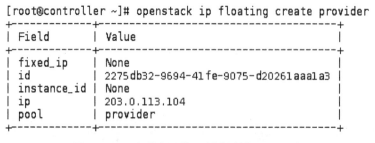

图 15 -42　在私有网络上创建浮动 IP 地址池

3）为实例分配浮动 IP

openstack ip floating add 203.0.113.104 selfservice-instance

如图 15 -43 所示。

```
[root@controller ~]# openstack ip floating add 203.0.113.104 selfservice-instance
[root@controller ~]#
```

图 15 - 43　为实例分配浮动 IP

4) 检查这个浮动 IP 地址的状态

openstack server list

如图 15 - 44 所示。

```
[root@controller ~]# openstack server list
+--------------------------------------+---------------------+--------+----------------------------+
| ID                                   | Name                | Status | Networks                   |
+--------------------------------------+---------------------+--------+----------------------------+
| fb9841eb-daf5-4b5b-                  | selfservice-instance | ACTIVE | selfservice=172.16.1.3,   |
| be74-3cc051b05c67                    |                     |        | 203.0.113.104              |
| 0ab30d71-c566-4928-bb8c-             | provider-instance   | ACTIVE | provider=203.0.113.103     |
| 9c1fdd0eba15                         |                     |        |                            |
+--------------------------------------+---------------------+--------+----------------------------+
```

图 15 - 44　检查浮动 IP 状态

5) 在虚拟机实例控制台中，验证能否 ping 通浮动 IP 地址

$ ping-c 4 203.0.113.104

如图 15 - 45 所示。

```
$ ping -c 4 203.0.113.104
PING 203.0.113.104 (203.0.113.104): 56 data bytes
64 bytes from 203.0.113.104: seq=0 ttl=63 time=1.689 ms
64 bytes from 203.0.113.104: seq=1 ttl=63 time=2.176 ms
64 bytes from 203.0.113.104: seq=2 ttl=63 time=1.648 ms
64 bytes from 203.0.113.104: seq=3 ttl=63 time=1.677 ms

--- 203.0.113.104 ping statistics ---
4 packets transmitted, 4 packets received, 0% packet loss
round-trip min/avg/max = 1.648/1.797/2.176 ms
```

图 15 - 45　控制台验证能否 ping 通浮动 IP 地址

6) 验证控制节点或者其他私有网络上的主机通过浮动 IP 地址 ping 通实例

ping-c 4 203.0.113.104

如图 15 - 46 所示。

```
[root@controller ~]# ping -c 4 203.0.113.104
PING 203.0.113.104 (203.0.113.104) 56(84) bytes of data.
64 bytes from 203.0.113.104: icmp_seq=1 ttl=63 time=3.74 ms
64 bytes from 203.0.113.104: icmp_seq=2 ttl=63 time=0.764 ms
64 bytes from 203.0.113.104: icmp_seq=3 ttl=63 time=0.794 ms
64 bytes from 203.0.113.104: icmp_seq=4 ttl=63 time=0.600 ms

--- 203.0.113.104 ping statistics ---
4 packets transmitted, 4 received, 0% packet loss, time 3002ms
rtt min/avg/max/mdev = 0.600/1.476/3.749/1.314 ms
```

图 15 - 46　从控制节点或私有网络主机进行 ping 验证

7）在控制节点或其他私有网络上的主机使用 SSH 远程访问实例

ssh cirros@203.0.113.104

如图 15-47 所示。

```
[root@controller ~]# ssh cirros@203.0.113.104
The authenticity of host '203.0.113.104 (203.0.113.104)' can't be established.
RSA key fingerprint is 1b:9e:f7:4d:e1:15:f3:0e:b9:68:c8:d7:c7:b0:ab:e2.
Are you sure you want to continue connecting (yes/no)? yes
Warning: Permanently added '203.0.113.104' (RSA) to the list of known hosts.
$
$ exit
Connection to 203.0.113.104 closed.
```

图 15-47　使用 SSH 远程访问实例

SSH 到实例时不需要另行输入用户登录密码。

15.6　创建块设备存储（控制节点）

在块存储服务上可以创建一个卷设备并连接到一个虚拟机实例上。

15.6.1　创建卷设备

1）获得 demo 凭证来获取只有非管理员才能执行的命令的访问权限

./root/demo-openrc

2）创建一个 1GB 的卷设备

openstack volume create--size 1 volume1

如图 15-48 所示。

```
[root@controller ~]# openstack volume create --size 1 volume1
+---------------------+--------------------------------------+
| Field               | Value                                |
+---------------------+--------------------------------------+
| attachments         | []                                   |
| availability_zone   | nova                                 |
| bootable            | false                                |
| consistencygroup_id | None                                 |
| created_at          | 2017-04-05T08:12:24.754482           |
| description         | None                                 |
| encrypted           | False                                |
| id                  | e7892f60-7291-425b-88bc-c197ee329428 |
| multiattach         | False                                |
| name                | volume1                              |
| properties          |                                      |
| replication_status  | disabled                             |
| size                | 1                                    |
| snapshot_id         | None                                 |
| source_volid        | None                                 |
| status              | creating                             |
| type                | None                                 |
| updated_at          | None                                 |
| user_id             | 455de29c0a3a4b86ac22b9f3f0b8ec67     |
+---------------------+--------------------------------------+
```

图 15-48　创建一个 1GB 的卷设备

3）列出可用的卷设备

\# openstack volume list

如图 15-49 所示。

```
[root@controller ~]# openstack volume list
+--------------------------------------+--------------+-----------+------+-------------+
| ID                                   | Display Name | Status    | Size | Attached to |
+--------------------------------------+--------------+-----------+------+-------------+
| e7892f60-7291-425b-88bc-c197ee329428 | volume1      | available |   1  |             |
+--------------------------------------+--------------+-----------+------+-------------+
```

图 15-49　列出可用的卷设备

当创建过程完全成功后，状态会从 creating 变为 available。

15.6.2　分配卷设备

分配卷设备给虚拟机实例：

1）获得 demo 凭证来获取只有非管理员才能执行的命令的访问权限

\# ./root/demo-openrc

2）创建一个 1 GB 的卷设备

将卷设备 volume1 分配给虚拟机实例 selfservice-instance。

\# openstack server add volume selfservice-instance volume1

如图 15-50 所示。

```
[root@controller ~]# openstack server add volume selfservice-instance volume1
[root@controller ~]#
```

图 15-50　创建一个 1 GB 的卷设备

3）列出可用的卷设备

\# openstack volume list

如图 15-51 所示。

```
[root@controller ~]# openstack volume list
+--------------------------------------+--------------+--------+------+----------------------------------+
| ID                                   | Display Name | Status | Size | Attached to                      |
+--------------------------------------+--------------+--------+------+----------------------------------+
| e7892f60-7291-425b-88bc-             | volume1      | in-use |   1  | Attached to selfservice-         |
| c197ee329428                         |              |        |      | instance on /dev/vdb             |
+--------------------------------------+--------------+--------+------+----------------------------------+
```

图 15-51　列出可用的卷设备

4）在虚拟机实例控制台中，使用 fdisk 命令验证 /dev/vdb 块存储设备已作为卷存在于系统中

\# sudo fdisk-l

如图 15-52 所示。

```
$ sudo fdisk -l
Disk /dev/vda: 1073 MB, 1073741824 bytes
255 heads, 63 sectors/track, 130 cylinders, total 2097152 sectors
Units = sectors of 1 * 512 = 512 bytes
Sector size (logical/physical): 512 bytes / 512 bytes
I/O size (minimum/optimal): 512 bytes / 512 bytes
Disk identifier: 0x00000000

   Device Boot      Start         End      Blocks   Id  System
/dev/vda1   *       16065     2088449     1036192+  83  Linux

Disk /dev/vdb: 1073 MB, 1073741824 bytes
16 heads, 63 sectors/track, 2080 cylinders, total 2097152 sectors
Units = sectors of 1 * 512 = 512 bytes
Sector size (logical/physical): 512 bytes / 512 bytes
I/O size (minimum/optimal): 512 bytes / 512 bytes
Disk identifier: 0x00000000

Disk /dev/vdb doesn't contain a valid partition table
```

图 15-52　使用 fdisk 命令验证 /dev/vdb 块存储设备已作为卷存在于系统中

只有在 /dev/vdb 设备上创建文件系统并挂载它，才能使用这个卷。

15.7　创建编排（控制节点）

编排服务可以创建一个栈来自动化创建一个实例。

15.7.1　创建模板

编排服务使用模版来描述栈。

创建 /root/demo-template.yml 文件：

gedit /root/demo-template.yml

添加以下内容：

heat_template_version: 2015-10-15

description: Launch a basic instance with CirrOS image using the "m1.nano" flavor, "mykey" key, and one network.

parameters:
　NetID:
　　type: string
　　description: Network ID to use for the instance.

resources:
　server:

```
type: OS::Nova::Server
    properties:
        image: cirros
        flavor: m1.nano
        key_name: mykey
        networks:
            - network: { get_param: NetID }

outputs:
    instance_name:
        description: Name of the instance.
        value: { get_attr: [ server, name ] }
    instance_ip:
        description: IP address of the instance.
        value: { get_attr: [ server, first_address ] }
```

15.7.2 创建栈

使用 demo-template.yml 模版创建一个栈。

1) 获得 demo 凭证来获取只有非管理员才能执行的命令的访问权限

\# . /root/demo-openrc

2) 列出可用的网络

\# openstack network list

如图 15-53 所示。

```
[root@controller ~]# openstack network list
+--------------------------------------+-------------+--------------------------------------+
| ID                                   | Name        | Subnets                              |
+--------------------------------------+-------------+--------------------------------------+
| e57af6ca-6f43-4aba-8d74-bc9ef537eea9 | provider    | e2c834a7-c2f5-4b5c-abae-4bd0f0177604 |
| c97a05d9-b771-49e7-884e-559270aba95b | selfservice | cb4abfb4-dfc8-4725-8b8d-807bd536cb4f |
+--------------------------------------+-------------+--------------------------------------+
```

图 15-53 列出可用网络

在使用网络时，必须使用的是网络 ID，而不是网络的名称。

3) 设置 NET_ID 环境变量表示网络 ID 例如：使用私有网络。

\# export NET_ID = $ (openstack network list | awk '/ selfservice / { print $ 2 }')

如图 15-54 所示。

```
[root@controller ~]# export NET_ID=$(openstack network list | awk '/ selfservice / { print $2 }')
[root@controller ~]#
```

图 15-54　设置 NET_ID 环境变量表示网络 ID

4) 在私有网络上创建一个 CirrOS 实例的栈

openstack stack create -t /root/demo-template.yml \
　　--parameter "NetID=$ NET_ID" stack

如图 15-55 所示。

```
[root@controller ~]# openstack stack create -t /root/demo-template.yml \
> --parameter "NetID=$NET_ID" stack
+---------------------+-----------------------------------------------------------+
| Field               | Value                                                     |
+---------------------+-----------------------------------------------------------+
| id                  | defa5cc3-e6ec-4d58-9c8e-0c39f1d27c86                      |
| stack_name          | stack                                                     |
| description         | Launch a basic instance with CirrOS image using the ``m1.nano`` flavor, |
|                     | ``mykey`` key, and one network.                           |
| creation_time       | 2017-04-05T08:24:54                                       |
| updated_time        | None                                                      |
| stack_status        | CREATE_IN_PROGRESS                                        |
| stack_status_reason | Stack CREATE started                                      |
+---------------------+-----------------------------------------------------------+
```

图 15-55　在私有网络上创建一个 CirrOS 实例的栈

5) 列出模板栈

openstack stack list

如图 15-56 所示。

```
[root@controller ~]# openstack stack list
+--------------------------------------+------------+-----------------+----------------------+--------------+
| ID                                   | Stack Name | Stack Status    | Creation Time        | Updated Time |
+--------------------------------------+------------+-----------------+----------------------+--------------+
| defa5cc3-e6ec-4d58-9c8e-0c39f1d27c86 | stack      | CREATE_COMPLETE | 2017-04-05T08:24:54  | None         |
+--------------------------------------+------------+-----------------+----------------------+--------------+
```

图 15-56　列出模板栈

执行完成后，状态会从 CREATE_IN_PROGRESS 变为 CREATE_COMPLETE。

6) 查看实例的名称和 IP 地址并和 OpenStack client 的输出作比较

openstack stack output show--all stack

如图 15-57 所示。

```
[root@controller ~]# openstack stack output show --all stack
+---------------+----------------------------------------------------------+
| Field         | Value                                                    |
+---------------+----------------------------------------------------------+
| instance_name | {                                                        |
|               |     "output_value": "stack-server-qdq2dhkaj526",         |
|               |     "output_key": "instance_name",                       |
|               |     "description": "Name of the instance."               |
|               | }                                                        |
| instance_ip   | {                                                        |
|               |     "output_value": "172.16.1.4",                        |
|               |     "output_key": "instance_ip",                         |
|               |     "description": "IP address of the instance."         |
|               | }                                                        |
+---------------+----------------------------------------------------------+
```

图 15 – 57 查看实例的名称和 IP 地址并和 OpenStack client 的输出作比较

7）检查虚拟机实例的状态

openstack server list

如图 15 – 58 所示。

```
[root@controller ~]# openstack server list
+--------------------------------------+---------------------------+--------+-------------------------------+
| ID                                   | Name                      | Status | Networks                      |
+--------------------------------------+---------------------------+--------+-------------------------------+
| 7cd3c21a-8eb4-400e-bfc8-e835636c683e | stack-server-qdq2dhkaj526 | ACTIVE | selfservice=172.16.1.4        |
| fb9841eb-daf5-4b5b-be74-3cc051b05c67 | selfservice-instance      | ACTIVE | selfservice=172.16.1.3,       |
|                                      |                           |        | 203.0.113.104                 |
| 0ab30d71-c566-4928-bb8c-9c1fdd0eba15 | provider-instance         | ACTIVE | provider=203.0.113.103        |
+--------------------------------------+---------------------------+--------+-------------------------------+
```

图 15 – 58 检查虚拟机实例的状态

8）删除 stack

openstack stack delete--yes stack

如图 15 – 59 所示。

```
[root@controller ~]# openstack stack delete --yes stack
[root@controller ~]#
```

图 15 – 59 删除 stack

9）再次检查虚拟机实例的状态

openstack server list

如图 15 – 60 所示。

```
[root@controller ~]# openstack server list
+--------------------------------------+--------------------+--------+---------------------------+
| ID                                   | Name               | Status | Networks                  |
+--------------------------------------+--------------------+--------+---------------------------+
| fb9841eb-daf5-4b5b-                  | selfservice-instance | ACTIVE | selfservice=172.16.1.3, |
| be74-3cc051b05c67                    |                    |        | 203.0.113.104             |
| 0ab30d71-c566-4928-bb8c-             | provider-instance  | ACTIVE | provider=203.0.113.103    |
| 9c1fdd0eba15                         |                    |        |                           |
+--------------------------------------+--------------------+--------+---------------------------+
```

图 15-60 再次检查虚拟机实例的状态

15.8 访问仪表板（控制节点）

通过仪表板可以查看、配置和管理 OpenStack 云平台组件。例如可以通过仪表板查看本章此前创建的虚拟机实例相关资源的配置信息。

在 Web 浏览器中，访问仪表板地址为 http://controller/dashboard，如图 15-61 所示。

图 15-61 在 Web 浏览器中访问仪表板

在前面的步骤中，已通过命令行（CLI）的方式创建了网络、主机类型和实例等内容。在仪表板 Web 界面，可以查看此前创建的这些内容，同时它还支持创建、修改和删除等操作。

管理员（例如 admin）和非管理员（例如 demo）的管理选项是不同的。

当以管理员身份登录时，仪表板可提供项目、管理员、身份管理选项，如图 15 – 62 所示。

图 15 – 62　仪表板的管理员界面

当以非管理员身份登录时，仪表板只提供项目和身份管理选项，如图 15 – 63 所示。

图 15 – 63　仪表板的非管理员界面

15.8.1　身份管理页面

当以管理员身份登录时，在管理员的身份管理页面中，可提供项目、用户、组和角色选项。

项目（租户）是 OpenStack 项目中的一个组织单元，用户可以同时属于一个或多个项目（租户），角色定义了用户可以执行的操作类型。具有管理权限的管理员可以管理所有的项目（租户）、用户和角色。

项目（租户）、用户和角色彼此之间没有依赖性，可以分别进行操作。在建立项目时，至少要创建一个项目（租户）、用户和角色。在进行删除用户的操作时，需要首先将用户和租户的映射删除。

项目页面展示了项目（租户）信息，并可对其进行配置管理操作，如图 15-64 所示。

图 15-64　项目页面

用户页面展示了用户信息，并可对其进行配置管理操作，其中包含了用以访问 OpenStack 各组件服务 API 端点的服务用户，如图 15-65 所示。

图 15-65　用户页面

组页面展示了组信息，并可对其进行配置管理操作，如图 15-66 所示。

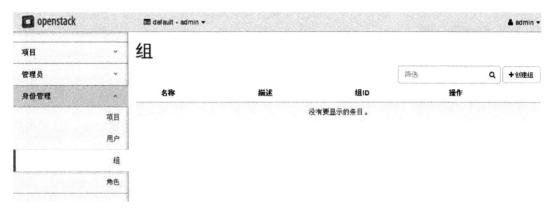

图 15-66 组页面

角色页面展示了角色信息,并可对其进行配置管理操作,如图 15-67 所示。

图 15-67 角色页面

当以非管理员身份登录时,身份认证仅可查看用户所属项目(租户)的项目信息,且不可以进行配置管理操作,如图 15-68 所示。

图 15-68 非管理员身份认证页面

15.8.2 管理员页面

仅当以管理员身份登录时，才会显示管理员选项。

在管理员选项中，展示了与 OpenStack 云平台相关的所有系统信息，并可在其中对各管理选项进行相应的配置管理操作。

管理选项包括概况、资源使用量、虚拟机管理器、主机聚合、云主机、卷、云主机类型、镜像、网络、路由、默认值、元数据定义和系统信息。

概况页面展示了 OpenStack 云平台使用情况的概况摘要，如图 15-69 所示。

图 15-69　概况页面

资源使用量页面展示了 OpenStack 云平台的资源使用概况（明细报表和数据），如图 15-70 所示。

图 15-70　资源使用量页面

虚拟机管理器页面展示了 OpenStack 云平台虚拟机管理器资源使用的总体概况，并提供了虚拟机管理器与计算主机明细信息，如图 15-71 所示。

图 15-71　虚拟机管理器页面

主机聚合页面展示了 OpenStack 云平台计算主机聚合信息，并可对其进行配置管理操作，如图 15-72 所示。

图 15-72　主机聚合页面

云主机页面展示了 OpenStack 云平台云主机（虚拟机实例）信息，并可对其进行配置管理操作，如图 15-73 所示。

图 15-73　云主机页面

卷页面展示了块存储服务的卷挂载信息，并可对其进行配置管理操作，如图 15-74 所示。

图 15-74　卷页面

云主机类型页面展示了云主机类型（虚拟机硬件配置）信息，并可对其进行配置管理操作，如图 15-75 所示。

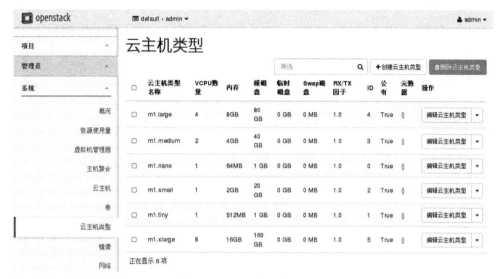

图 15-75　云主机类型页面

镜像页面展示了镜像信息，并可对其进行配置管理操作，云主机需依赖于镜像，如图 15-76 所示。

图 15-76　镜像页面

网络页面展示了网络信息，并可对其进行配置管理操作，云主机需依赖于网络，如图 15-77 所示。

图 15-77　网络页面

路由页面展示了路由信息，并可对其进行配置管理操作，如图 15 - 78 所示。

图 15 - 78　路由页面

默认值页面展示了 OpenStack 平台的配额默认值信息，并可对其进行配置管理操作，如图 15 - 79 所示。

图 15 - 79　默认值页面

系统信息页面展示了 OpenStack 平台系统服务和各组件服务的详细信息，如图 15 - 80 至 15 - 84 所示。

图 15-80　系统信息页面 1（服务信息）

图 15-81　系统信息页面 2（计算服务信息）

图 15-82　系统信息页面 3（块存储服务信息）

图 15-83　系统信息页面 4（网络代理信息）

图 15-84　系统信息页面 5（Orchestration 服务）

15.8.3　项目页面

项目仅展示当前登录用户所用资源的信息，包括计算、网络、云编排以及对象存储。默认情况下，各项目（租户）的资源是不可相互访问的。

概况页面，展示了资源使用的概况信息。

本书示例中，由于 admin 用户没有创建任务云主机（虚拟机实例），因此，计算和存储资源概况图表展示是空的，如图 15-85 所示。

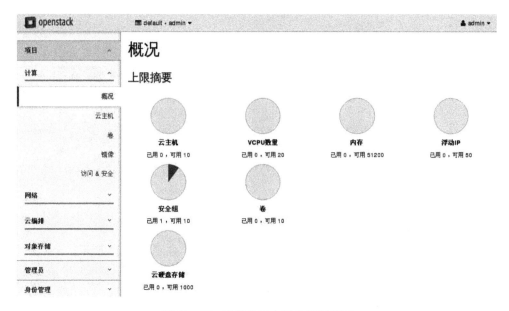

图 15-85　无任务云主机的概况页面

本书示例中,由于 demo 用户创建了两个任务云主机(虚拟机实例),因此,资源概况图表展示了相应的使用量,如图 15-86 所示。

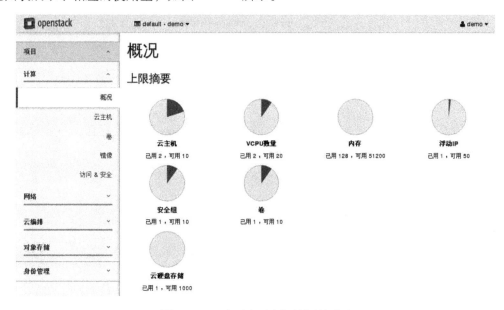

图 15-86　有两个云主机的概况页面

云主机页面展示了云主机(虚拟机实例)信息,并可对其进行配置管理操作,如图 15-87 所示。

图 15-87 云主机页面

卷页面展示了卷信息,并可对其进行配置管理操作,如图 15-88 所示。

图 15-88 卷页面

访问 & 安全页面展示了安全、密钥对、浮动 IP 和访问 API 端点信息,并可对其进行配置管理操作,如图 15-89 至 15-92 所示。

图 15-89 访问 & 安全页面 1(安全组信息)

图 15-89　访问 & 安全页面 2（密钥对信息）

图 15-91　访问 & 安全页面 3（浮动 IP 信息）

图 15-92　访问 & 安全页面 4（访问 API 信息）

网络拓扑页面展示了云主机的网络拓扑图示结构信息，并可对其进行配置管理操作，如图 15-93 所示。

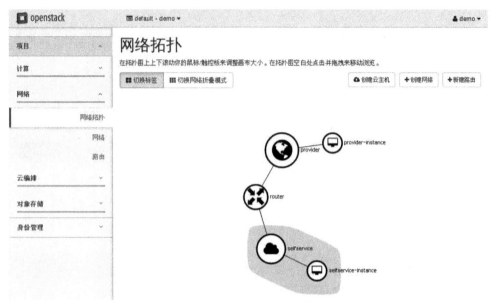

图 15-93　网络拓扑页面

网络页面展示了网络信息，并可对其进行配置管理操作，如图 15-94 所示。

图 15-94　网络页面

路由页面展示了路由信息，并可对其进行配置管理操作，如图 15-95 所示。

图 15-95　路由页面

参 考 文 献

[1] OpenStack Installation Guide for Red Hat Enterprise Linux and CentOS [EB/OL]. [2017-04-10] https://docs.openstack.org/mitaka/install-guide-rdo/.
[2] 卢万龙, 周萌. OpenStack 从零开始学 [M]. 北京：电子工业出版社, 2016.
[3] 沈建国, 陈永. OpenStack 云计算基础架构平台技术与应用 [M]. 北京：人民邮电出版社, 2017.